# Modern Small Arms

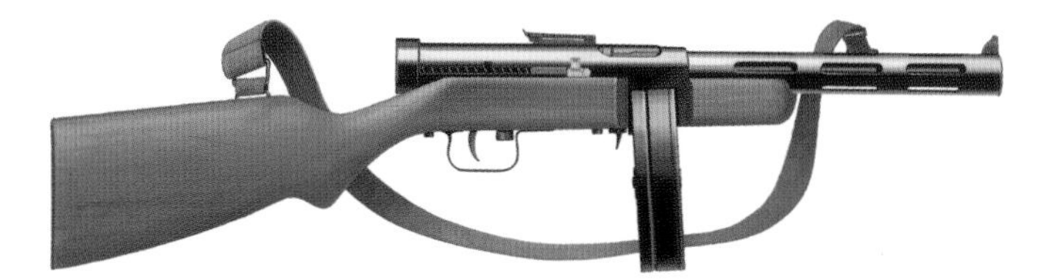

# Modern Small Arms

## 300 OF THE WORLD'S GREATEST SMALL ARMS

CHRIS McNAB

Reprinted in 2018, 2020

This new edition published in 2015



First published in 2001



Published by
Amber Books Ltd
United House
North Road
London N7 9DP
United Kingdom
www.amberbooks.co.uk
Instagram: amberbooksltd
Facebook: amberbooks
Twitter: @amberbooks

ISBN: 978-1-78274-216-6

Project Editor: Charles Catton
Editor: Siobhan O'Connor
Design: David Stanley

Printed in China

Picture credits:
**TRH Pictures**
Artwork credits: **Aerospace, Amber Books Ltd, John Batchelor, Chrysalis Picture Library, Istituto Geografico De Agostini S.p.A., Bob Garwood, Mainline Design (Guy Smith), Jan Suermondt.**

# CONTENTS

# Introduction

**The true revolution in weapons design must reside in the 19th century. Johann Nikolaus Dreyse's Zündnadelgewehr (needle gun) and Alphonse Chassepot's Fusil d'Infanterie Modèle 1866 established the basic design of the bolt-action rifle, while later in the century designers such as the Frenchman Basile Gras, the Austrian Ferdinand Mannlicher and the German Peter Paul Mauser would ally the bolt-action with magazine loading systems. Hand-cranked machine guns such as the Mitrailleuse and the Gatling, the latter with a 3000 rounds per minute (rpm) rate of fire, were already showing their destructive possibilities. More significantly, landmark figures, such as Hiram Stevens Maxim and Baron Odkolek von Augezd respectively, applied to machine guns the principles of recoil and gas operation which are used to this day. The Maxim gun, in particular, spread around the world and proved its hideous efficiency when used by the British in Africa and Afghanistan in the 1890s.**

In the world of the handgun, the 19th century was also a time of exceptional progress. After Samuel Colt brought his percussion revolver onto the market in 1835, and Horace Smith and Daniel B. Wesson introduced the first cartridge revolver, the pistol became a viable combat weapon. Double-action (meaning that the gun is cocked and fired in a single pull of the trigger), ejection rod emptying of cylinder cases and side-opening cylinders quickly followed; barring metallurgical limits and

*Japanese troops on the Great Wall of China. The officer is armed with a Mauser C96 pistol, while the machine gunner is ready to fire his Type 91.*

some technical sophistications, the revolvers of the late 19th century could claim a capability little different from the handguns of today. In addition, designers such as Hugo Bochardt were already introducing the automatic, magazine-loaded pistol – a form that would overtake the revolver in popularity in the 20th century.

Combine the progress in gun design with the allied fact that the 19th century gave us the true unitary cartridge and more efficient nitro-cellulose propellants, and the 20th century may seem to play second fiddle. However, whereas the 19th century was perhaps the age of the greater experimentation, the 20th was the century in which experiment would give way to excellence.

## TESTED IN FIRE

World War I can be seen as the point in history when the old ways of individuated war were swept away by the industrialised destruction brought by new firearms technologies. Effective bolt-action rifles were now in the hands of all soldiers, and a two-man machine-gun team could deliver the firepower previously held by an entire company. The smokeless propellants now used converted almost all their explosive energy into gas, thus giving rifles the high velocities needed to kill at ranges of 1000m- (3280ft-) plus, while remaining hidden. Firepower was now superior to manpower, as was illustrated when, at Mons in August 1914, the 7500 men of the British Expeditionary Force, armed with Lee-Enfield rifles and limited numbers of Vickers machine guns, stopped the advance of 200,000 soldiers of the German 1st Army.

The consequence of these technologies was that troop exposure on open ground became suicidal. Thus, military tactics shifted to become more defensive in nature, employing groundworks, trenches and fortifications. Although such defensive tactics had already been seen in the South African War (1899–1902) and the Russo-Japanese War (1904–1905), World War I shifted them to a previously unimaginable scale. As so often happens in weapons development, the conditions imposed by new firearms created the need for new tactics, which in turn produced new weapons to support those tactics. By 1916, German forces were employing *Sturmtruppe* – small groups of heavily armed shock troops that would puncture sections of enemy trenches following a short, intense artillery bombardment. In the confines of trenches, wielding a 1.25m- (49.2in-) long Mauser Gewehr 98 bolt-action rifle was awkward, its five-round magazine was inadequate and its long-range power was unwarranted. Two new types of small arm were born – the submachine gun and the light machine gun.

The submachine gun answered the needs of trench warfare by providing full automatic fire using pistol-calibre cartridges, usually the ubiquitous 9mm

Parabellum, to make the recoil manageable, and had a convenient length and weight for the trench confines. Range was limited – about 30–40m (98–131ft) – but more than adequate for most actual combat distances. The first submachine gun was Italian: the Vilar-Perosa was a blowback weapon set in a double-barrel configuration with a cyclic rate of fire of 1200rpm. Although designed for Austrian Alpine troops, it was actually too heavy for convenient personal use and was more suited to being mounted on vehicles and in aircraft. By contrast, the German Bergmann MP18 was a reliable, powerful 9mm weapon, purposely designed for trench-clearing operations. It was an excellent gun and proved popular with the German shock troops; and the submachine gun would become a fundamental type of infantry weapon for the next three decades.

The light machine gun originated out of a similar impulse towards more mobile firepower. Heavy machine guns such as the Vickers, Maxim and Schwarzlose were incredibly weighty when fully armed and assembled on their tripod unit. A Maxim ready for firing could weigh around 62kg (137lb) which, although light enough to be manhandled by a team, still lacked the portability to act as a forward-support weapon during assaults. The answer was the light machine gun, introduced from about 1915. Light enough to be carried by one man, these guns were usually gas-operated, air-cooled weapons which could be quickly set up and fired from a bipod. For convenience, they tended to use a magazine rather than belt feed. Significant examples were the British Lewis gun, the US M1918 Browning Automatic rifle, the Hotchkiss Mle 1909, the German MG'08/15, and such weapons dramatically increased the weight of fire that could be brought to bear during assaults.

The pistol was also a natural favourite for trench combat, and, by World War I, pistol development was centring on the automatic pistol. Classics such as John Browning's infamous M1911 and George Luger's Pistole '08 had already entered service, and self-loading pistols by Beretta, Savage, Webley & Scott, and Mauser were making an impact. Usually operating by blowback or short recoil, automatic pistols had the advantage of holding more rounds than a revolver's cylinder and, usually, offered a more convenient loading method through a detachable magazine or a charger of rounds inserted into an internal magazine.

The end of World War I left military strategists with many important tactical lessons, not least of which was the greater value of short-range portable firepower for infantry combat than long-range guns. Thus the interwar period continued the development of the light- and submachine gun form, although the bolt-action rifle would continue as the main infantry weapon for most nations until at least the end of World War II (Japan, in particular, lagged behind).

The interwar period and, later, World War II did, however, take the technologies of the machine gun in all its forms to new heights. To list the wartime submachine guns – weapons such as the German MP38/40, the British Sten, the Russian PPSh41, and the US M3 and Thompson – is to perform a roll call for some of the most famous firearms ever invented. Yet the conditions of mass mobile warfare imposed by German blitzkrieg meant that war had to be fought on the production lines as well as on the battlefields. Thus industrialised production methods using steel pressings and stampings became integral to any successful weapon design in World War II. The Soviet PPSh41, for example, was simple enough to produce even in small rural workshops and consequently it was possible to keep Soviet forces stocked with this invaluable source of firepower.

Major advances in production and sophistication of machine guns were also taking place. While some, such as the estimable Browning M2HB, were making reputations that would keep them in use for the next half century, World War II was the period in which the light machine gun reached its apotheosis. Two German examples, in particular – the MG34 and MG42 – demonstrated such high rates of fire, dependability, range and general killing force that they questioned the relevance of heavy machine guns at all. The success of such weapons in mobile support roles was such that many are with us today in forms such as the German M3.

World War II was a watershed in 20th-century weapons design not only for the quality (and quantity) of machine gun technology which emerged, but also for the introduction of a new weapon form that remains, to this day, the central infantry firearm of all major armies – the self-loading automatic rifle. Although the Russian Vladimir Fyodorov had created a 6.5mm automatic rifle during World War II, it was the US M1 Garand that established itself as the first significant self-loading rifle, although it retained the full rifle calibre .30-06 round. The true assault rifle was designed from the recognition that the standard high-power rifles of the day had a range far in excess of the average 400m (1297ft) distance at which men fought. In response, the visionary Hugo Schmeisser designed the Sturmgewehr (assault rifle) 44 in 1941. Looking ultra-modern for the time, this was a gas-operated weapon the long, curved magazine of which held thirty 7.92 x 33mm 'Kurz' (Short) cartridges which were the same calibre as Mauser rifle rounds, but with reduced power for more range-realistic usage and controllability at full automatic.

The Schmeisser assault rifle came too late in the war to make a difference and fewer than 450,000 were made. Yet, with the end of World War II, most major powers were starting to recognise that the assault rifle would become the best infantry weapon, situated as it was between the pistol-calibre, short-range

*Men of the US 96th Division on Okinawa in 1945, armed with the M1 Garand (left) and Browning Automatic Rifle (right).*

submachine gun and the long-range rifle. In 1947, the most famous assault rifle of all time was produced by Mikhail Timofeyevich Kalashnikov – the AK-47. It fired a 7.62mm intermediate cartridge and impressed the world with its ability to deliver heavy individual firepower while enduring the worst battlefield conditions.

Although it would later catch up, the newly formed NATO missed a bold chance to match the AK-47's useful cartridge with something similar by agreeing to standardise all NATO weapons to 7.62 x 51mm NATO. This round was over-powerful for an assault rifle with automatic capabilities, but certain US parties were reluctant to give up the long-range round. Many fine weapons were produced for this calibre – the US M14, the superb FN FAL and Heckler & Koch's G3 – but they all struggled under fully automatic fire.

The solution had begun to emerge from the USA shortly after the Korean War. Research had begun into small-calibre high-velocity (1000mps/3200fps) ammunition, specifically the 5.56mm round. A rifle to fire this round was found in Eugene Stoner's AR-15, what would become the M16 rifle after it was adopted by the US Air Force in the early 1960s. The M16 used a highly efficient gas-operation and could fire easily on full automatic. Furthermore, the velocity of the small round was such that it retained the stopping and killing power of much larger rounds by force of its supersonic shock effect.

The 5.56mm round was resisted for many years (its cause not helped by early jamming problems with the M16), yet, after the US Army adopted it for general use in the M16A1 in Vietnam, the direction was inexorable. In the late 1970s, the Soviet Union brought out a 5.45mm version of the AK-47, the AK-74; after trials in the 1980s, the 5.56 x 45mm round was adopted as the NATO standard. New weapons emerged. Heckler & Koch brought out the G41, the FN FAL rifle became the FN FNC and Israel produced the Galil. 'Bullpup' designs also gained currency. These were weapons that located much of the receiver behind the trigger unit and thus were able to maximise the length of the barrel for accuracy, while restraining the overall length of the weapon. Some of these designs – particularly the British Enfield L85A1, the French FAMAS and the Austrian Steyr AUG – have become standard-issue infantry weapons in some armies.

While assault rifles have generally eclipsed submachine guns in their prominence following World War II, submachine gun technology and designs also proceeded apace. Perhaps the greatest development is in the area of compaction. High-quality submachine guns such as the Heckler & Koch MP5 series retain more rifle-like proportions, but other submachine guns have become little bigger than pistols. By siting the magazine within the pistol grip and using a telescoping bolt (i.e. a bolt which actually encloses the end of the barrel), guns such as the Uzi, the Ingram M10 and the South African BXP are easily concealed, yet can spray out devastating firepower in close-combat situations. At the other end of the scale, standard assault rifles have also been developed with longer, heavier barrels to become squad automatic weapons (SAWs). These are intended to give small infantry units greater sustained-fire capability and range with their standard calibre weapons and from standard magazines. The jury is still out as to whether they perform a valuable function over the general-purpose machine gun (GPMG) type that has been used as a heavier support weapon by most armies since World War II.

The astonishing scientific progress in terms of ballistics, materials and manufacturing in the 20th century means that even the humble pistol is a work of mastery. Modern handguns such as the SIG-Sauer P226 have 15-round magazines or, like the Glock pistol range, a higher percentage of their build in plastic than in metal. Sniper rifles have gone even further. The contemporary sniper armed with, say, an FR-F1 or an L96A1 and looking through an advanced-optics telescopic sight can confidently expect a first-round kill at 800m (2624ft), while, in the Gulf War, a sniper armed with a Barrett .50in rifle took a confirmed kill at 1800m (5905ft).

The big question is, what next? In terms of weapons that use conventional ammunition and methods of operation, we have perhaps gone as far as we can.

*The current French assault rifle is the FAMAS, a bullpup design. Here it is being fired on the range by members of the French Foreign Legion.*

Recent experiments have been conducted using machine guns that fire bullets through electromagnetic acceleration rather than percussion, the result being an utterly silent yet astonishingly dense and powerful rain of high-velocity fire. Other ideas are already off the drawing board. Heckler & Koch's G11 rifle fires caseless ammunition in which the bullet is embedded in a rectangle of propellant which disappears completely on firing. The removal of the need for ejection gives the rifle a very high rate of fire – its three-round burst sounds as a single explosion.

As is so often the way for military equipment, politics often restrain gun development. Thus, the G11 was passed over as the next German Army rifle, the 7.62mm NATO round remained in service for more years than necessary, and legal actions are still being taken against the US Army for adopting the Beretta 92 instead of US-built firearms in the 1980s. However, as history has shown, progress in weapons design is inexorable, and the next stage will soon be with us.

NOTE: The entries are arranged chronologically by country within each sub-type.

# Mannlicher Model 1901

**The Model 1901 was the first of a series of Mannlicher firearms produced by the well-known Austrian manufacturer Steyr between 1901 and 1905, although the first Mannlicher pistol was actually produced in 1900 by Von Dreyse. All the pistols were quality weapons, even if both their commercial and their military success were limited. Two features distinguish the M1901. First, the delayed-blowback operation worked on a system in which the delay was imposed by a spring-and-cam system restraining the slide during its rearward travel. Secondly, the Mannlicher guns have an integral magazine which is loaded by pushing a charger of cartridges down through the open slide. One of the biggest markets for the guns was the Argentine Army, and the 7.63mm ammunition still remains in production in South America today.**

| | |
|---|---|
| Country of origin: | Austria |
| Calibre: | 7.63 x 21mm Mannlicher |
| Length: | 246mm (9.68in) |
| Weight: | 0.91kg (2lb) |
| Barrel: | 157mm (6.18in), 4 grooves, rh |
| Feed/magazine capacity: | 8-round integral box magazine |
| Operation: | Delayed blowback |
| Muzzle velocity: | 312mps (1024fps) |
| Effective range: | 30m (98ft) |

# Steyr Model 1917

**The Steyr M1917 was a later variation of the estimable Steyr M1912 pistol, one of the finest handguns of the first half of the 20th century. The M1912 was the standard Imperial Austro-Hungarian (later Austrian) service pistol between 1912 and 1945, and, in common with other Austrian weapons, was also used by German forces during World War II (although in the latter case the gun was recalibrated for the ubiquitous German 9mm Parabellum round, rather than the more powerful 9mm Steyr). Known as the 'Steyr-Hahn' ('Steyr-Hammer') because of its external hammer, the M1912 had a rotating barrel for breech locking, the barrel turning 20° on firing before the slide disengaged and completed its recoil. Loading the Steyr's integral magazine consisted of inserting a charger of bullets down into the pistol's fixed magazine.**

| | |
|---|---|
| **Country of origin:** | Austria |
| **Calibre:** | 9mm Steyr or 9mm Parabellum |
| **Length:** | 216mm (8.5in) |
| **Weight:** | 0.99kg (2.19lb) |
| **Barrel:** | 128mm (5in), 4 grooves, rh |
| **Feed/magazine capacity:** | 8-round integral box magazine |
| **Operation:** | Short recoil |
| **Muzzle velocity:** | 335mps (1100fps) |
| **Effective range:** | 30m (98ft) |

# Glock 17

**A remarkable weapon in terms of construction, materials and marketing, the Glock 17 has become a dominant force in the military and commercial handgun industries since its introduction. First produced in 1983, it demonstrates a 40 per cent use of plastic materials (although the barrel and slide naturally remain metal) and an inventory of only 33 parts for each gun. While the plastic makes it a light weapon, the Glock 17 also has an impressive magaxine capacity of no less than 17 rounds and fires its ammunition using a trigger-controlled striker instead of a hammer. It uses a locked breech with tilting barrel mechanism and safety is provided by a trigger safety and firing-pin lock. Through good marketing, and following its adoption by the Austrian Army, the Glock 17 has gone on to extensive military and police use across the world.**

| | |
|---|---|
| **Country of origin:** | Austria |
| **Calibre:** | 9mm Parabellum |
| **Length:** | 188mm (7.4in) |
| **Weight:** | 0.65kg (1.44lb) |
| **Barrel:** | 114mm (4.49in), 6 grooves, rh |
| **Feed/magazine capacity:** | 17-round detachable box magazine |
| **Operation:** | Short recoil |
| **Muzzle velocity:** | 350mps (1148fps) |
| **Effective range:** | 40m (131ft) |

# Glock 18

While, in 1986, Beretta were bringing out their fully automatic pistol, the Beretta 93R, Glock were also introducing their Model 18 handgun. This featured two extended magazine options – one with a 19-round capacity, the other holding as many as 33 rounds – and looked and worked very much like the Model 17 pistol. Yet, because of some internal modifications, the Glock 18 could fire at a high cyclic rate of around 1300rpm. Naturally, this fully automatic capability has taken the Glock 18 out of the commercial market in favour of government and military use. Whether these markets will find the gun practical only time will tell, but the rationale behind the fully automatic pistol, as opposed to simply having a small submachine gun, is questionable, as the former is difficult to control even if fired by an experienced shooter.

| | |
|---|---|
| Country of Origin: | Austria |
| Calibre: | 9mm Parabellum |
| Length: | 223mm (8.78in) |
| Weight: | 0.636kg (1.4lb) |
| Barrel: | 114mm (4.49in), hexagonal, rh |
| Feed/Magazine capacity: | 19- or 33-round detachable box magazine |
| Operation: | Short recoil |
| Muzzle Velocity: | 350mps (1148fps) |
| Effective Range: | 40m (131ft) |

# Glock 20

From the early 1990s, a new range of Glock pistols appeared on the market in larger calibres than the traditional 9mm Parabellum. The Glock 20 and 21 were available in 10mm Auto, an especially powerful pistol round to be handled only by the well-trained shooter used to handling such weapons (it soon became an FBI handgun). Glock was amongst the first gun manufacturers to make a 10mm gun, but the Glock 20 retained all the features of earlier weapons: a two-part trigger for effective safety control; a large capacity (15-round) magazine; and an overall light weight (0.78kg/1.7lb) owing to its predominantly plastic construction. The Glock 20 was followed by another 10mm gun, the Glock 21, after which nos. 22 and 23 arrived in .40 Smith & Wesson calibre. A new version of the weapon is now being produced in .45 ACP.

| | |
|---|---|
| Country of origin: | Austria |
| Calibre: | 10mm Auto |
| Length: | 193mm (7.59in) |
| Weight: | 0.78kg (1.73lb) |
| Barrel: | 117mm (4.6in), hexagonal, rh |
| Feed/magazine capacity: | 15-round detachable box magazine |
| Operation: | Short recoil |
| Muzzle velocity: | 350mps (1148fps) |
| Effective range: | 40m (131ft) |

# Browning Modèle 1900

**The long-standing association between John M. Browning and the Belgian arms manufacturer Fabrique Nationale d'Armes de Guerre (FN) led to the production of many classic firearms in the 20th century. The first of these was the Browning Modèle 1900, which was conceived in the late 19th century and went into service in 1900. It was a blowback pistol which set itself apart by having a recoil spring sited above the barrel that acted not only in recoil action, but also as the firing pin spring. The gun was an all-round success in terms of design, and some one million were produced in total. These were much used by military forces across the world – although usually without offical status, as the turn of the century was still a period of adjustment to the idea of the automatic pistol, and revolvers were seen as being more reliable weapons.**

| | |
|---|---|
| **Country of origin:** | Belgium |
| **Calibre:** | 7.65mm Browning (.32 ACP) |
| **Length:** | 170mm (6.75in) |
| **Weight:** | 0.62kg (1.37lb) |
| **Barrel:** | 101mm (4in), 6 grooves, rh |
| **Feed/magazine capacity:** | 7-round detachable box magazine |
| **Operation:** | Blowback |
| **Muzzle velocity:** | 290mps (950fps) |
| **Effective range:** | 30m (98ft) |

# Browning High-Power Model 1935

In many senses, the Browning High-Power laid the groundwork for the modern combat handgun. Designed in the 1920s by John M. Browning and produced at Fabrique Nationale d'Armes de Guerre (FN) in Belgium from 1935, the High-Power featured a 13-round capacity, double-row magazine (nearly twice the capacity of its main rival, the Colt M1911) and a different locking mechanism to the Colt, one which used a shaped cam to draw the barrel downwards to lock in a more linear fashion than the Colt's swinging link. The result was a reliable, accurate side arm which armed men from both sides during World War II and went on to serve in the armed forces of more than 50 countries to this day. Variations in the 1980s kept the High-Power relevant to the market and, even though the design is starting to date somewhat, it remains internationally popular.

| | |
|---|---|
| Country of origin: | Belgium/USA |
| Calibre: | 9mm Parabellum |
| Length: | 197mm (7.75in) |
| Weight: | 0.99kg (2.19lb) |
| Barrel: | 118mm (4.65in), 4 grooves, rh |
| Feed/magazine capacity: | 13-round detachable box |
| Operation: | Short recoil |
| Muzzle velocity: | 335mps (1100fps) |
| Effective range: | 30m (98ft) |

# Browning Double-Action

The Browning Double-Action (DA) is an advancement on the Browning High-Power, the main improvement being an ambidextrous de-cocking lever in place of a safety catch. This allows the user to drop the hammer safely, even when a round has been loaded into the chamber. However, the double-action mechanism means that, although the gun is completely safe in this mode, simply pulling the trigger through will fire the round. Fabrique Nationale d'Armes de Guerre (FN) continued the ambidextrous features – the magazine release switch, usually fitted on the right side, can be reversed for left-handed use. Apart from these features and a resculpting of the contours of the grip for two-handed use, the DA remains the same basic handgun as the High-Power, and its quality has made it a worthy replacement for the older weapon.

| | |
|---|---|
| **Country of origin:** | Belgium |
| **Calibre:** | 9mm Parabellum |
| **Length:** | 200mm (7.87in) |
| **Weight:** | 0.905kg (1.99lb) |
| **Barrel:** | 118mm (4.65in), 6 grooves, rh |
| **Feed/magazine capacity:** | 14-round detachable box |
| **Operation:** | Short recoil |
| **Muzzle velocity:** | 350mps (1148fps) |
| **Effective range:** | 30m (98ft) |

# CZ75

Since it first appeared in 1975, the CZ75 has excelled as both a combat and a commerical pistol, and has been much copied (made easier by the manufacturer's lack of effective patent cover outside Eastern Europe). It is based on the Colt-Browning dropping barrel action and has a double-action trigger mechanism. In terms of its action, there is little unusual to note, but the overall workmanship of the gun is high, and so the weapon gives the reliability and consistent performance required by military and police personnel. Ironically, the Czech military did not adopt the pistol (mainly because they followed Soviet calibrations), but many other world armies have taken the gun into service, either in its original format or as a copy (licensed or otherwise). A new model, the CZ85, has improved on the safety features.

| | |
|---|---|
| Country of origin: | Czechoslovakia |
| Calibre: | 9mm Parabellum |
| Length: | 203mm (8in) |
| Weight: | 0.98kg (2.16lb) |
| Barrel: | 120mm (4.72in), 6 grooves, rh |
| Feed/magazine capacity: | 15-round detachable box magazine |
| Operation: | Short recoil |
| Muzzle velocity: | 338mps (1110fps) |
| Effective range: | 40m (131ft) |

# Tokagypt 58

The ubiquity of the 9mm Parabellum cartridge in modern pistols formed the basis of this Egyptian weapon, which is essentially a copy of the Soviet Tokarev TT-33 pistol for the 9mm round. 'Copy' is an accurate term, for, apart from the butt grip and finish, there is little to distinguish it from the Soviet original in either action or performance (hence the hybrid name, Toka-gypt). Yet the change to a much more practical calibre has made a difference to its performance, and the gun was a decent weapon which ultimately had an inauspicious record of use. It was actually built in Hungary during the 1960s, but its intended user, the Egyptian Army, finally decided against the weapon as an issue firearm. The surplus stock was thus disseminated throughout the Egyptian police and also to various commercial parties in Western Europe.

| | |
|---|---|
| Country of origin: | Egypt/Hungary |
| Calibre: | 9mm Parabellum |
| Length: | 194mm (7.65in) |
| Weight: | 0.91kg (2.01lb) |
| Barrel: | 114mm (4.5in), 6 grooves, rh |
| Feed/magazine capacity: | 7-round detachable box magazine |
| Operation: | Short recoil |
| Muzzle velocity: | 350mps (1150fps) |
| Effective range: | 30m (98ft) |

# Helwan

The Helwan was a direct copy of Beretta's excellent Model M951 pistol for the Egyptian armed forces during the latter's build up; it was made under licence during the 1960s, chosen for service ahead of the Tokagypt. Being a direct copy, to describe its properties is to describe those of the original Beretta pistol, and the only way to distinguish the Helwan from the Beretta is by the inscription 'HELWAN CAL 9 m/m U.A.R.' on the weapon's slide. The Beretta M951/Helwan is a short-recoil pistol in which the locking of barrel and breech is achieved by locking lugs dropping into slots in the walls of the slide. Although Beretta originally, and unsuccessfully, attempted to produce the M951 in a light alloy, the final steel gun was a solid first foray into a 9mm weapon for Beretta, and Israel also adopted the design for service.

| | |
|---|---|
| Country of origin: | Egypt |
| Calibre: | 9 x 19mm Parabellum |
| Length: | 203mm (8in) |
| Weight: | 0.89kg (1.96lb) |
| Barrel: | 114mm (4.5in), 6 grooves, rh |
| Feed/magazine capacity: | 8-round detachable box magazine |
| Operation: | Short recoil |
| Muzzle velocity: | 350mps (1148fps) |
| Effective range: | 40m (131ft) |

# Chamelot-Delvigne 1874

The Chamelot-Delvigne was a Belgian weapon designed by Joseph Camelot and Henri-Gustave Delvigne which went on to become a standard French army pistol during the latter years of the 19th and early years of the 20th century. The series of guns actually emerged in 1865 but it was not until a new version was produced in 1871 that export sales increased. In 1873 an 11mm version was produced which was taken into service by the French cavalry. Early versions of this weapon were hampered by a chronically under-powered cartridge which limited what was a solid and reliable gun. This problem was eventually cleared with a new cartridge which offered nearly double the power, and for this round the 1874/90 version was produced. In this form, the Chamelot-Delvigne served French and other European troops into World War I.

| **Country of origin:** | France/Belgium |
| --- | --- |
| Calibre: | 11mm |
| Length: | 240mm (9.4in) |
| Weight: | 1.08kg (2.3lb) |
| Barrel: | 114mm (4.49in) |
| Feed/magazine capacity: | 6-round cylinder |
| Operation: | Revolver |
| Muzzle velocity: | 183mps (600fps) |
| Effective range: | 20m (66ft) |

# Manhurhin MR73

**A French-designed revolver of excellent overall quality, the Manhurhin MR73 is a versatile pistol which is available in a variety of calibres and barrel lengths to suit different shooting requirements, both private and military. The engineering of elements such as the trigger system – which is very smooth, owing to a separate spring – and barrel, which is cold-hammered, is especially fine. In fact the weapon possesses the facility for changing the .38 Special cylinder for a 9mm Parabellum cylinder in just a couple of minutes, giving the gun greater flexibility than the average revolver. Whether in these calibres or in the more common .357in Magnum, the Manhurhin MR73 is an accurate and stable weapon to fire, even when fitted with the shorter barrels favoured by the police and military for ease of concealment and lighter weight.**

| | |
|---|---|
| Country of origin: | France |
| Calibre: | .357in Magnum; .38 Special; 9mm Parabellum |
| Length: | 195mm (7.67in) |
| Weight: | 0.88kg (1.94lb) |
| Barrel: | 63.5mm (2.5in) |
| Cylinder capacity: | 6 rounds |
| Operation: | Revolver |
| Muzzle velocity: | According to cartridge |
| Effective range: | According to cartridge |

# MAB PA-15

MAB stands for Manufacture d'Armes de Bayonne, a long-standing French weapons producer who supplied the French forces with a standard service pistol, the PA-15, for more than two decades. The PA-15 was visually stylish and worked its way through a number of calibres, first starting with the 7.65mm ACP and 9mm Short cartridges, and then adopting the standard 9mm Parabellum round. In this latter calibration, the PA-15 had to work on a delayed-blowback system of operation using a rotating barrel. The initial forces of recoil hold both slide and barrel locked in place through a grooved track until the bullet has left the muzzle, then pressures drop and the barrel turns, allowing the slide to recoil. The PA-15 gave good service in the French Army between 1975 and 1990 (but its manufacturer, MAB, no longer exists).

| | |
|---|---|
| **Country of origin:** | France |
| **Calibre:** | 9 x 19mm Parabellum |
| **Length:** | 203mm (8in) |
| **Weight:** | 1.07kg (2.36lb) |
| **Barrel:** | 114mm (4.5in), 6 grooves, rh |
| **Feed/magazine capacity:** | 15-round detachable box magazine |
| **Operation:** | Delayed blowback |
| **Muzzle velocity:** | 330mps (1100fps) |
| **Effective range:** | 40m (131ft) |

# Mauser Zig-Zag

The Mauser Zig-Zag was one of a series of Mauser weapons which emerged during the latter years of the 19th century and influenced the weapons of the 20th by either service or design. The most noticeable feature of these guns were their mechanism for revolving the cylinder during firing. This consisted of a set of grooves machined into the outer wall of the cylinder, into which a stud attached to the mainspring carrier was located. Thus, as the trigger was pulled, the movement of the mainspring carrier and stud was directed into the grooves and the cylinder revolved round to the next chamber for firing. The positioning of the grooves ensured that the chamber was correctly aligned for each firing. Mauser Zig-Zags came in solid and hinge-frame versions, with some (as pictured here) having the feature of simultaneous cartridge ejection via a rod located beneath the barrel.

| | |
|---|---|
| Country of origin: | Germany |
| Calibre: | 10.9mm |
| Length: | 298mm (11.75in) |
| Weight: | 1.19kg (2.6lb) |
| Barrel: | 165mm (6.5in) |
| Feed/magazine capacity: | 6-round cylinder |
| Operation: | Revolver |
| Muzzle velocity: | 198mps (650fps) |
| Effective range: | 30m (98ft) |

# Bochardt C/93

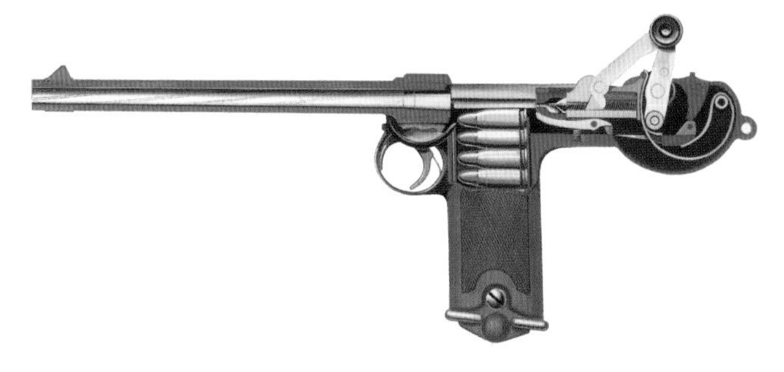

**Produced as it was at the end of the 19th century, the Bochardt C/93 was a significant step forwards towards an effective self-loading pistol, though it was quickly surpassed by the likes of Mauser and Bergmann. It was a large and heavy weapon with a distinctive 'fishing reel' appearance on account of its toggle-lock system of locking. This was borrowed from the Maxim machine gun, and the C/93 used a similar short-recoil principle of operation. Perhaps the most significant advance in the Bochardt handgun was that it went from charger loading into a fixed magazine to using a detachable 8-round box magazine, which was inserted in the bottom of the pistol grip in the way that has now become familiar. The C/93's production ceased in 1898, but it served into the twentieth century, often fitted with a shoulder stock to allow it to act as a carbine.**

| | |
|---|---|
| Country of origin: | Germany |
| Calibre: | 7.65mm |
| Length: | 279mm (11in) |
| Weight: | 1.1kg (2.56lb) |
| Barrel: | 165.1mm (6.5in), 4 grooves, rh |
| Feed/magazine capacity: | 8-round detachable box magazine |
| Operation: | Short recoil |
| Muzzle velocity: | 326mps (1070fps) |
| Effective range: | 20m (65.6ft) |

# Bergmann 1896

Bergmann started producing automatic pistols in 1894 in a variety of calibres including 5mm, 6.5mm and 8mm. These experiments were a little precarious to say the least, owing to an extraction method which involved the spent case ricocheting off the new round sitting behind it. The 1896, however, used a more standard extractor, and thus had enough success to keep it in work into the twentieth century. It was lighter in weight than many other self-loading pistols of the time, and was also compact enough for genuine useability. Loading to the integral magazine was done via an access plate on the magazine which was released by catch just in front of the trigger guard. Bergmann quickly superseded the 1896 model handgun with new designs and went on to produce some seminal firearms during the 20th century.

| | |
|---|---|
| Country of origin: | Germany |
| Calibre: | 7.63mm |
| Length: | 245mm (9.6in) |
| Weight: | 1.13kg (2.5lb) |
| Barrel: | 102mm (4in) |
| Feed/magazine capacity: | 5-round integral box magazine |
| Operation: | Blowback |
| Muzzle velocity: | 380mps (1250fps) |
| Effective range: | 30m (98ft) |

# Bergmann-Bayard M1910

**Not to be confused with the Mauser C/96 which it superficially resembles, the Bergmann-Bayard was developed by Bergmann in the early 1900s, but mainly manufactured by the Pieper company, who purchased the rights from Bergmann in 1908. The takeover was necessary, as Bergmann had received an order for his pistol from the Spanish Army which he could not meet. Pieper also standardised the weapon to 9mm, which had been Bergmann's dominant choice, even though he did experiment with other calibres such as 10mm and .45in. The Bergmann-Bayard was a heavy though solid weapon with a short-recoil mechanism which was essentially reliable. Apart from the Spanish Army, Greek and Danish forces also accepted the weapon as a standard side arm, and it continued in service until the end of World War II.**

| | |
|---|---|
| **Country of origin:** | Germany/Denmark |
| **Calibre:** | 9 x 23mm Bergmann-Bayard |
| **Length:** | 254mm (10in) |
| **Weight:** | 1.02kg (2.25lb) |
| **Barrel:** | 101mm (4in), 6 grooves, lh |
| **Operation:** | Short recoil |
| **Feed/magazine capacity:** | 10-round detachable box magazine |
| **Muzzle velocity:** | 305mps (1000fps) |
| **Effective range:** | 30m (98ft) |

# Mauser C/12

**The Mauser C/12, or M1912, was the main military version of the famous 'broomhandle' C/96, which was designed by the three Feederle brothers and went into production with Mauser in 1896. Many variations followed, including models fitted with shoulder stocks, but the military model arrived in 1912 and is representative of most of the pistol series. Chambered for 7.63mm Mauser and, later in World War I, 9mm Parabellum, the C/96 was extremely well balanced despite its front-heavy appearance and was diligently machined. Its operation was a complex short-recoil and its integral magazine was top-loaded with chargers. This magazine was varied between 6-, 10- and 20-round versions, the latter proving impractical and soon being dropped. In short, the Mauser pistols were high-performance weapons but their downfall was their high cost of manufacture.**

| | |
|---|---|
| Country of origin: | Germany |
| Calibre: | 7.63mm Mauser or 9mm Parabellum |
| Length: | 312mm (12.25in) |
| Weight: | 1.25kg (2.75lb) |
| Barrel: | 139mm (5.5in), 4 or 6 grooves, rh |
| Feed/magazine capacity: | 10-round integral box magazine |
| Operation: | Short recoil |
| Muzzle velocity: | 434mps (1425fps) |
| Effective range: | 60m (196ft) |

# Parabellum P'08 (Luger)

The Parabellum pistol's ancestry reaches back into the 19th century, but it is probably defined by the Pistole '08, known after its designer, Georg Luger. The transition from the 7.65mm calibre of Luger's early guns to 9mm Parabellum secured the popularity of the Pistole '08, and it was adopted by both the German navy and army. More than 2.5 million were subsequently produced between 1908 and 1945. Ironically, its visually distinctive toggle-lock mechanism was both a virtue and a problem – a virtue in that it worked well, but a problem in that it only worked well if kept clean, something far from easy in combat conditions. Nevertheless, it fired accurately (especially when fitted with an optional shoulder stock), was comfortable for the firer to hold and, even when simplified for wartime manufacture, exuded quality.

| | |
|---|---|
| Country of origin: | Germany |
| Calibre: | 9mm Parabellum |
| Length: | 233mm (8.75in) |
| Weight: | 0.87kg (1.92lb) |
| Barrel: | 102mm (4in), 6 grooves, rh |
| Feed/magazine capacity: | 8-round detachable box |
| Operation: | Short recoil |
| Muzzle velocity: | 380mps (1247fps) |
| Effective range: | 30m (98ft) |

# Parabellum Artillery Model

**The Parabellum Artillery Model (also known as the 'Long '08') was an early attempt to transform a standard pistol into a more potent carbine for use at longer ranges. In this case, it was simply achieved by extending the barrel of a standard '08 pistol to 190mm (7.5in), adding a leaf sight at the rear of the barrel, and supplying a wooden shoulder stock for attachment at the base of the pistol grip. In addition, the gun would also take the 32-round 'snail' magazine like that used by the Bergmann MP18, as well as the standard 8-round box magazine. More than 140,000 Artillery Models were made, the intention being to supply artillerymen, engineers and aircrew, who, it was thought, would need a greater range. However, overall the design lacked some plausibility and was never as popular in service as the standard pistol.**

| | |
|---|---|
| Country of origin: | Germany |
| Calibre: | 9mm Parabellum |
| Length: | 311mm (12.4in) |
| Weight: | 1.05kg (2.31lb) |
| Barrel: | 190mm (7.5in), 4 grooves, rh |
| Feed/magazine capacity: | 8-round box magazine or 32-round snail magazine |
| Operation: | Short recoil |
| Muzzle velocity: | 380mps (1250fps) |
| Effective range: | 40m (131ft) |

# Walther PPK

Although made famous by being the standard firearm of the fictional James Bond in Ian Fleming's books, the Walther PPK, although small, is a fairly unexceptional firearm. It emerged in 1931 as a smaller version of the PP pistol design for police use. As a double-action blowback weapon, it was reliable and light, although its varied calibres (.22 LR; 6.35mm Browning; 7.65mm Browning; 9mm Short) are somewhat indecisive when it comes to putting an opponent down. As a police weapon, the PPK has become dated, especially in its limited magazine capacity of only seven rounds, compared to the 15 available in a Beretta Model 92 or a SIG-Sauer P-225. However, its ease of portability and reliability mean that it will remain in service for many years to come amongst police and security units, who appreciate the ease with which it can be concealed.

| | |
|---|---|
| Country of origin: | Germany |
| Calibre: | .22 LR; 6.35mm or 7.65mm Browning (.32 ACP); 9mm Short |
| Length: | 148mm (5.8in) |
| Weight: | 0.59kg (1.3lb) |
| Barrel: | 80mm (3.15in), 6 grooves, rh |
| Feed/magazine capacity: | 7-round detachable box magazine |
| Operation: | Blowback |
| Muzzle velocity: | 290mps (950fps) |
| Effective range: | 30m (98ft) |

# Mauser HSc

The Mauser HSc started its production life in 1937. Like the Walther PP, it offered a double-action weapon of 7.65mm calibre – this was quite progressive for the Mauser company at the time, which was working in competition with the advanced Walther designers. Certain features, however, did distinguish it from the Walther. Even more of the hammer was concealed in the slide, with just a tiny portion left for cocking, and the whole gun was of a very sleek and minimalist design. The safety system actually disengaged the firing pin from its track into a recess that took it out of alignment with the cartridge. Production of the HSc (the initials standing for 'Hammerless, Self-Loading, Model C') ceased in 1945, but resumed commercially again in 1964. In the mid-1980s, Mauser licensed the HSc to the Italian arms company Renato Gamba.

| | |
|---|---|
| Country of origin: | Germany |
| Calibre: | 7.65mm Browning (.32 ACP) |
| Length: | 152mm (6in) |
| Weight: | 0.64kg (1.32lb) |
| Barrel: | 86mm (3.38in), 6 grooves, rh |
| Feed/magazine capacity: | 8-round detachable box magazine |
| Operation: | Blowback |
| Muzzle velocity: | 290mps (960fps) |
| Effective range: | 30m (98ft) |

# Walther P38

Part of Germany's armaments rationalisation and expansion in the 1930s was the request for a new service pistol to replace the Luger P'08. Walther set to the task, modifying its PP pistol and going through various formats until the 9mm 'HP' (*Heeres Pistole*) was accepted, this then being designated the Pistole 38, or P38. The qualities of the P38 would see it in service throughout the 1950s and into the present day (known after 1957 as the P1). Well made, attractively plated in matt black and very reliable, the P38 featured a safety indicator pin which showed whether there was a cartridge in the chamber or not, and it also had an advanced double-action lock, which enabled the operator to fire the weapon from a hammer-down position with just a single pull on the trigger. It stands as one of the 20th century's best handguns.

| | |
|---|---|
| Country of origin: | Germany |
| Calibre: | 9mm Parabellum |
| Length: | 213mm (8.38in) |
| Weight: | 0.96kg (2.11lb) |
| Barrel: | 127mm (5in), 6 grooves, rh |
| Feed/magazine capacity: | 8-round detachable box |
| Operation: | Short recoil |
| Muzzle velocity: | 350mps (1150fps) |
| Effective range: | 30m (98ft) |

# Heckler & Koch P9

**The Heckler & Koch (H&K) P9 comes in two versions: the standard P9 which is a single-action pistol (operation of the internal hammer is achieved by a release and cocking lever on the left side of the frame) and the P9S, which is a double-action weapon. What both have in common, however, is that they use the H&K roller-locked delayed-blowback system used in the G3 assault rifle. Thus, as recoil drives the bolt system rearwards, two rollers lock into barrel extensions and hold the bolt until pressure is at a safe level. Both the H&K P9 and PS9 amount to fine handguns, used by several police and military units around the world; .45in calibre (for the US market) and 7.65mm Parabellum versions have also been issued. One item of note is that the bore has a polygonal configuration, with the rifling grooves set into the bore diameter.**

| | |
|---|---|
| **Country of origin:** | Germany |
| **Calibre:** | 9mm Parabellum or .45 ACP |
| **Length:** | 192mm (7.56in) |
| **Weight:** | 0.88kg (1.94lb) |
| **Barrel:** | 102mm (4in), polygonal, rh |
| **Feed/magazine capacity:** | 9-round detachable box (9mm); 7-rnd detach. box (.45 ACP) |
| **Operation:** | Roller-locked delayed blowback |
| **Muzzle velocity:** | 350mps (1150fps) |
| **Effective range:** | 40m (131ft) |

# Heckler & Koch P7

The P7 was purpose-designed for the Federal German Police by Heckler & Koch as a very safe but rapid-response combat weapon. Two features distinguish this gun from other security handguns: the cocking lever and the mode of operation. The cocking lever is situated just in front of the grip. When out, it entirely prohibits the firing of the gun even if there is a round in the chamber. To fire, the user must squeeze in the grip to cock the firing pin before pulling the trigger. This gives a weapon which is very safe when not in use, but which can quickly be brought into action when need be. The operation is gas-actuated delayed blowback, in which gases are channelled from the barrel to momentarily resist the slide's recoil until the round has left the muzzle. This excellent, well-designed and popular weapon has entered broad global use.

| | |
|---|---|
| Country of origin: | Germany |
| Calibre: | 9mm Parabellum |
| Length: | 171mm (6.73in) |
| Weight: | 0.8kg (1.76lb) |
| Barrel: | 105mm (4.13in), polygonal, rh |
| Feed/magazine capacity: | 13-round detachable box magazine |
| Operation: | Gas-actuated delayed blowback |
| Muzzle velocity: | 350mps (1150fps) |
| Effective range: | 40m (131ft) |

# Walther P5

The Walther P5 was produced by the company in the 1970s to meet the demanding safety criteria of the West German police, who were looking for a new handgun. The gun's actual firing mechanism is essentially that of the excellent Walther P38, but with three specific safety features. First, only if the trigger is pulled will the firing pin strike the cartridge, as usually the firing pin sits in a recess in the hammer – pulling the trigger realigns the firing pin with the impact portion of the hammer. Added to this is a safety notch for the hammer, and the fact that the gun will not fire unless the slide is in a fully closed position. In addition to its excellent safety precautions, the P5 is also an excellent weapon to fire, and its overall fine qualities have taken it into service outside Germany, in countries such as Portugal and Holland.

| | |
|---|---|
| Country of origin: | Germany |
| Calibre: | 9mm Parabellum |
| Length: | 180mm (7.08in) |
| Weight: | 0.79kg (1.75lb) |
| Barrel: | 90mm (3.54in), 6 grooves, rh |
| Feed/magazine capacity: | 8-round detachable box magazine |
| Operation: | Short recoil |
| Muzzle velocity: | 350mps (1150fps) |
| Effective range: | 40m (131ft) |

# Webley Bulldog

The Webley Bulldog was a stocky little revolver which was produced by Webley from the late 1870s, and whose durability as a firearm took it into the 20th century, if only in service rather than in production. It was developed from an earlier Webley revolver produced for the Royal Irish Constabulary, and first arrived on the scene in .442in calibre, then went to .45in, then to .32in. Whatever the calibre, each Bulldog could be relied upon to deliver workable firepower for its user from its five-chamber cylinder. The guns were also cheap, and a combination of good price and solidity made the Bulldog a popular weapon which became widely distributed. The main problem with the Bulldog was that its very short barrel – only 53mm (2.1in) – which gave it an acutely compressed effective range of around 15m (49ft).

| | |
|---|---|
| Country of origin: | Great Britain |
| Calibre: | .32in British |
| Length: | 140mm (5.5in) |
| Weight: | 0.31kg (0.7lb) |
| Barrel: | 53mm (2.1in) |
| Feed/magazine capacity: | 5-round cylinder |
| Operation: | Revolver |
| Muzzle velocity: | 190mps (625fps) |
| Effective range: | 15m (49ft) |

# Webley-Fosbery

The Webley-Fosbery's appearance as a conventional revolver belies its revolutionary, albeit inefficient, operation as a halfway house between a revolver and an automatic pistol. It was designed by Colonel G. V. Fosbery VC in the mid-1890s and it would stay in service with the British armed forces until the end of World War I. On firing the Webley-Fosbery, the barrel and cylinder would recoil along a slide on the top of the butt and trigger system. This would recock the hammer, while a stud on the slide engaged with grooves on the cylinder to turn the cylinder to the next round. Despite its ingenuity, the Webley-Fosbery was an unwieldy weapon which was hard to control on firing and vulnerable to dirt – something that was hardly in short supply on the Western Front. By 1915, production of it had ceased.

| | |
|---|---|
| **Country of origin:** | Great Britain |
| **Calibre:** | .455in British Service |
| **Length:** | 279mm (11in) |
| **Weight:** | 1.25kg (1.378lb) |
| **Barrel:** | 152mm (6in), 7 grooves, rh |
| **Cylinder capacity:** | 6 rounds |
| **Operation:** | Automatic revolver |
| **Muzzle velocity:** | 183mps (600fps) |
| **Effective range:** | 30m (98ft) |

# Webley & Scott Mk 6

**The Mk VI was the last of a long line of redoutable Webley service revolvers produced from the late 1880s to the end of WWII. The Mk VI was officially in service between 1915 and 1945, and continued the Webley pistol's reputation for being heavy, accurate and reliable. It was also powerful. Firing the large .455 cartridge required a strong arm, but as a man-stopper it was undeniably successful and was used to great effect in close-quarters trench combat in WWI. A Pitcher/Greener revolver bayonet was also available, though not in popular use, as it made the gun even more unwieldy. One of the virtues of all Webley service revolvers was that they could be used under the filthiest of conditions without giving cause for concern, and their durability means that many collector's models are still being fired today.**

| | |
|---|---|
| Country of origin: | Great Britain |
| Calibre: | .455 British Service |
| Length: | 286mm (11.25in) |
| Weight: | 1.09kg (2.4lb) |
| Barrel: | 152mm (6in), 7 grooves, rh |
| Cylinder capacity: | 6-rounds |
| Operation: | Revolver |
| Muzzle velocity: | 200mps (655fps) |
| Effective range: | 30m (98ft) |

# Webley & Scott Self-Loading Pistol 1912 Mk 1

This unusual and heavy gun was an early foray by Webley & Scott into the world of the automatic pistol, although they themselves disliked this term and preferred 'self-loading'. A short-recoil weapon, the Mk 1 used the immensely powerful .445 W&S Auto cartridge, for many years the world's most potent pistol round. This round was not transferable with the .445 ammunition used by Webley revolvers, a point which several unfortunates discovered when revolver cylinders exploded during firing the Auto cartridge. The Mk 1 was issued to the Royal Navy, the Royal Flying Corps, elements of the Royal Horse Artillery and some British police units, but it was never as popular with its users as the Webley revolvers. The force of the gun and its awkward angularity made it somewhat difficult to fire in rapid-response situations.

| | |
|---|---|
| Country of origin: | Great Britain |
| Calibre: | .445in W&S Auto |
| Length: | 216mm (8.5in) |
| Weight: | 0.62kg (1.37lb) |
| Barrel: | 127mm (5in), 6 grooves, rh |
| Feed/magazine capacity: | 7-round detachable box magazine |
| Operation: | Short recoil |
| Muzzle velocity: | 228mps (750fps) |
| Effective range: | 30m (98ft) |

# Enfield .38

The Enfield pistol emerged from the demand for a less powerful British handgun that could be more easily handled by troops than the muscular .455 Webley pistols which served the British forces in WWI. A .38in 200-grain round was selected as appropriate for the role and the Royal Small Arms factory put their 'Pistol, Revolver, Number 2 Mark 1' into production in 1926/27. In most senses it was a copy of the Webley Mark VI, with modifications to the trigger and safety mechanisms. It was a double-action revolver, and this became the only method of operation when the Mark 1* was brought out with the hammer spur removed. Several different versions of the Enfield pistol were produced and it, along with the Webley Mk IV, became the standard issue service pistols for the British Army during WWII. Although efficient, they were not as prized as their German counterparts.

| | |
|---|---|
| Country of origin: | Great Britain |
| Calibre: | .38 British Service |
| Length: | 260mm (10.25in) |
| Weight: | 0.78kg (1.72lb) |
| Barrel: | 127mm (5in), 7 grooves, rh |
| Cylinder capacity: | 6 rounds |
| Operation: | Revolver |
| Muzzle velocity: | 198mps (650fps) |
| Effective range: | 30m (98ft) |

# Fromer Model 1910

Rudolf Frommer's Model 1910 was unusual for adopting a long-recoil principle of operation (where the barrel and bolt recoil for a distance greater than the entire length of the cartridge). The gun was actually developed some years earlier than its model date, but competition from the Roth-Steyr Model 1907 forced Frommer to improve the design before it could successfully enter the market. The principle was not ideal for a pistol, and though the Model 1910 was reliable enough, by 1930 Frommer was replacing it with a series of other handguns based on the Browning system of blowback operation (though not before another long-recoil pistol design, the Frommer 'Stop', was produced from 1912). Frommer's blowback guns proved to be simple, controllable firearms, and they became widely distributed throughout Europe.

| | |
|---|---|
| Country of origin: | Hungary |
| Calibre: | 7.65mm Browning |
| Length: | 184mm (7.25in) |
| Weight: | 0.63kg (1.43lb) |
| Barrel: | 100mm (4in), 4 grooves, rh |
| Feed/magazine capacity: | 7-round detachable box magazine |
| Operation: | Long recoil |
| Muzzle velocity: | 335mps (1099fps) |
| Effective range: | 20m (65ft) |

# IMI Desert Eagle

Part of a modern return to vogue of large-calibre combat pistols, the Desert Eagle originated in the USA, but went through to production at Israel Military Industries (IMI) in Israel. It comes in three main Magnum calibres – .357, .44 and .50 – all of which produce a conclusive stopping power. The mechanism of the Desert Eagle is gas-operated. On firing, gas is drawn off through a port just in front of the chamber; this pushes back the slide, which ejects the spent round from the chamber and then returns and reloads the weapon under the pressure of a return spring. The gas bleed reduces the significant recoil of the Desert Eagle to manageable levels for the firer, but it is still an incredibly powerful weapon – it can even come with extended barrels and telescopic sights – which has yet to persuade military circles of its validity.

| | |
|---|---|
| Country of origin: | Israel/USA |
| Calibre: | .357in, .44in or .50in Magnum |
| Length: | 260mm (10.25in) |
| Weight: | 1.7kg (3.75lb) .357; 1.8kg (4.1lb) .44; 2.05kg (4.5lb) .50 |
| Barrel: | 152mm (6in) |
| Feed/magazine capacity: | 9 rounds (.357 Mag.); 8 rounds (.44 Mag.); 7 rounds (.50 Mag.) |
| Operation: | Gas |
| Muzzle velocity: | 436mps (1430fps) for .357 Magnum; 448mps (1470fps) .44 Magnum |
| Effective range: | 50m (164ft) plus |

# Pistola Automatica Glisenti Modello 1910

Designed by Revelli and initially manufactured in 7.65mm calibre by Siderurgica Glisenti until 1907, the Glisenti gun was developed further by Metallurgica Brescia gia Tempini, which responded to the Italian Army's request for a 9mm calibre weapon in 1910. The Modello 1910's fundamental problems centred on its construction. The left side of the gun was designed to come almost completely away for ease of cleaning, a feature which simply weakened the gun's endurance. Consequently, a less powerful cartridge, the 9mm Glisenti, was developed with the same dimensions as the more powerful Parabellum. Mixing up the two, as happened, could cause a breech explosion dangerous to the user. A further model of the Glisenti with minor safety changes emerged (Modello 1912), but as a combat gun it had severe limitations.

| | |
|---|---|
| Country of origin: | Italy |
| Calibre: | 9mm Glisenti |
| Length: | 207mm (8.15in) |
| Weight: | 0.82kg (1.81lb) |
| Barrel: | 100mm (3.94in), 6 grooves, rh |
| Feed/magazine capacity: | 7-round detachable box |
| Operation: | Short recoil |
| Muzzle velocity: | 280mps (919fps) |
| Effective range: | 20m (66ft) |

# Beretta Modello 1934

The Modello 1934 was the standard Italian Army pistol between 1934 and 1945. It was actually part of a long evolution of Beretta automatic pistols, developing from the Modello 1915/19, although with an external hammer like the Modello 1931. If anything, the hammer was a flaw in the design, as it remained operable even when the safety was on and the trigger locked – meaning that users had to take care. Another irritant was that the slide snapped forwards when an empty magazine was taken out instead of waiting open to receive a fresh magazine; thus the operator had to carry out a lot of slide operation to reload. Despite these problems and the fact that the 9mm Short round did not carry quite enough power for combat use, it was a generally superb and reliable weapon which is much sought after by collectors today.

| | |
|---|---|
| Country of origin: | Italy |
| Calibre: | 9mm Short |
| Length: | 152mm (6in) |
| Weight: | 0.66kg (1.46lb) |
| Barrel: | 94mm (3.7in), 6 grooves, rh |
| Feed/magazine capacity: | 7-round detachable box magazine |
| Operation: | Blowback |
| Muzzle velocity: | 250mps (820fps) |
| Effective range: | 30m (98ft) |

# Beretta 81

The Beretta Model 81 was developed and brought into service in the mid 1970s, part of Beretta's move towards double-action handguns with a greater rate of fire. Two models were brought out at this time: the Model 81 in 7.65mm calibre and the Model 84 for 9mm Short ammunition. Despite its age, the Model 81 is a thoroughly modern pistol in its design. It has a 12-round magazine capacity, and its double-fire action allows the user to pull the trigger twice on the same round in the case of a first-attempt misfire. Safety for users is also improved. The pistol cannot be fired if the magazine has been removed from the gun, even if a round is still seated in the chamber. The extractor protrudes from the weapon and shows red if there is a round in the chamber and the safety switch can be easily reached by both left and right-handed shooters.

| | |
|---|---|
| Country of origin: | Italy |
| Calibre: | 7.65mm |
| Length: | 172mm (6.77in) |
| Weight: | 0.655kg (1.4lb) |
| Barrel: | 97mm (3.81in), 6 grooves, rh |
| Feed/magazine capacity: | 12-round detachable box magazine |
| Operation: | Blowback |
| Muzzle velocity: | 300mps (985fps) |
| Effective range: | 30m (98ft) |

# Beretta Model 93R

The Beretta 93R is an attempt to give a standard pistol greater capability in close-quarters combat by giving it a selective-fire facility. A switch on the side of the gun allows the user to choose between single shots and a three-round burst, the latter giving the weapon a cyclic rate of 1100rpm. Naturally, stability in the standard pistol configuration is compromised in such a mode, so the 93R features a folding foregrip just in front of the trigger that gives the automatic mode a genuine useability. The firer is able to control the weapon's recoil, meaning that the shots can be fired in a close grouping. Further control is added by the option of a folding metal stock and a short muzzle brake/flash hider. The 93R is used by some special forces in both Italy and abroad, but the concept of the pistol/machine pistol has yet to gain worldwide currency.

| | |
|---|---|
| Country of origin: | Italy |
| Calibre: | 9mm Parabellum |
| Length: | 240mm (9.45in) |
| Weight: | 1.12kg (2.47lb) |
| Barrel: | 156mm (6.14in), 6 grooves, rh |
| Feed/magazine capacity: | 15- or 20-round detachable box |
| Operation: | Short recoil |
| Muzzle velocity: | 375mps (1230fps) |
| Effective range: | 40m (131ft) |

# Beretta Model 92SB

The Beretta Model 92SB had the honour of winning the US Army's trials in the 1980s for a replacement side arm to the M1911, although some further modifications had to be made before it went into US service as the Model 92F. The 92SB is actually part of an extensive Model 92 series, all of superb quality and designed around a practical range of military and security needs. Origins of the 92 series began in an update of the Model 951 with a greater magazine capacity and a double action. Subsequent models improved on various elements and, by the time the 92SB reached the US trials, it had an ambidextrous safety and magazine catch and a half-cock facility. The 92F saw further modifications to satisfy US purchase demands, mainly limited to the gun's ergonomics, but also including the internal chroming of the barrel.

| | |
|---|---|
| Country of origin: | Italy |
| Calibre: | 9mm Parabellum |
| Length: | 197mm (7.76in) |
| Weight: | 0.98kg (2.16lb) |
| Barrel: | 109mm (4.29in), 6 grooves, rh |
| Feed/magazine capacity: | 13-round detachable box magazine |
| Operation: | Short recoil |
| Muzzle velocity: | 385mps (1263fps) |
| Effective range: | 40m (131ft) |

# Meiji Revolver

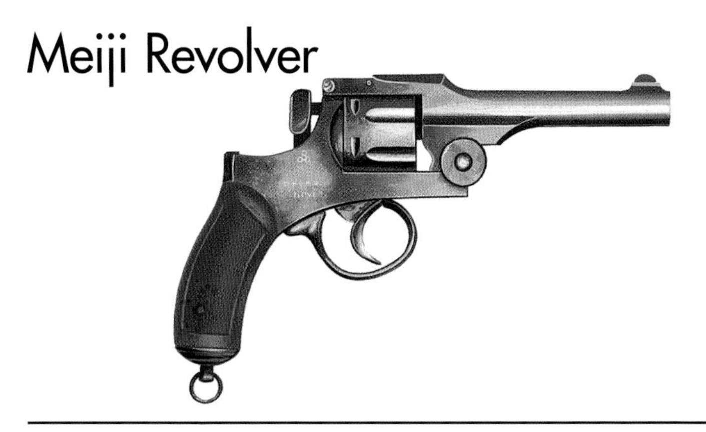

The Meiji revolver, or the Meiji 26 Nen Shiki Kenju, to give it its full name, was produced in the last years of the 19th century as part of Japan's general attempt to catch up with the rest of the Western industrialised world in terms of arms manufacture. Japan's relative inexperience in this field perhaps shows in the design, with borrowings from Smith & Wesson, Gasser and Nagant which seem to inform the pistol's construction. The Meiji was a double-action only revolver, and proved to be a rather inadequate gun for service, being both inaccurate and tending towards unreliability. It was .35in in calibre and used a break-open cylinder for reloading. Cartridge ejection was performed by an automatic ejector. The Meiji did not match the quality of the guns it imitated, yet it was still put into service in large numbers across Japan.

| | |
|---|---|
| Country of origin: | Japan |
| Calibre: | 9mm |
| Length: | 235mm (9.25in) |
| Weight: | 0.91kg (2lb) |
| Barrel: | 119mm (4.7in) |
| Feed/magazine capacity: | 6-round cylinder |
| Operation: | Revolver |
| Muzzle velocity: | 246mps (750fps) |
| Effective range: | 30m (91ft) |

# Nambu 14th Year

The Nambu 14 was in many regards the same design as the disappointing Nambu 1904 handgun, but was cheaper to manufacture. It entered the Japanese market in 1925 and was officially accepted by the Imperial Japanese Army in 1927. The main difference from the Nambu 1904 was a safety catch on the receiver and, in 1939, the trigger guard's dimensions were expanded to enable the gun to be fired with a gloved hand. These changes did not, however, make the gun a dependable combat tool for those unfortunate enough to be equipped with it. The Nambu 14's magazine retaining spring and two recoil springs were prone to failure in corrosive climates – the pressure they imparted could make it difficult for a soldier to reload his weapon with sweaty or wet hands, thus making the Nambu 14 a dangerous weapon on which to rely.

| | |
|---|---|
| Country of origin: | Japan |
| Calibre: | 8mm Nambu |
| Length: | 227mm (8.93in) |
| Weight: | 0.9kg (1.98lb) |
| Barrel: | 121mm (4.76in), 6 grooves, rh |
| Feed/magazine capacity: | 8-round detachable box |
| Operation: | Short recoil |
| Muzzle velocity: | 335mps (1100fps) |
| Effective range: | 30m (98ft) |

# Radom wz.35

The Radom wz.35 was designed in Poland in the 1930s in an attempt to create a standardised Polish army side arm. It saw service between 1936 and 1945, although for most of the war years it was produced under German occupation. This period saw the quality of the weapon reduced, as the Germans tried to speed up gun production for their own use. As its appearance suggests, the Radom was beholden to the Colt-Browning system. Of short-recoil operation, its substantial weight enabled it to handle the 9mm round comfortably and overall it was an excellent gun. Its one drawback was an inadequate safety system, as it had no safety mechanism (the switch on the left of the receiver is for use during stripping) apart from a grip safety, although it did have a system for lowering the hammer onto a loaded chamber without firing.

| | |
|---|---|
| Country of origin: | Poland |
| Calibre: | 9mm Parabellum |
| Length: | 197mm (7.76in) |
| Weight: | 1.022kg (2.25lb) |
| Barrel: | 115mm (4.53in), 6 grooves, rh |
| Feed/magazine capacity: | 8-round detachable box |
| Operation: | Short recoil |
| Muzzle velocity: | 350mps (1150fps) |
| Effective range: | 30m (98ft) |

# Unceta Victoria

The Unceta Victoria was produced in Spain in the first decade of the 1900s, and offered a compact handgun in a relatively light 7.65mm calibre. As a company, Esperanza y Unceta was founded in 1908, and became a prolific gun manufacturer, though nowadays it is better known as the manufacturer Astra, which it became later, named after the Unceta logo. The Unceta Victoria was one of a (lengthy) line of European turn-of-the-century guns which paid a visible homage to Browning's ground-breaking handgun designs, and the general robustness of the Victoria imitation saw it achieve adoption by the French army as a sidearm. This, however, was to end when the Victoria was generally superseded by the Campo Giro, the latter gun going on to become the Spanish army's standard sidearm, with some examples seeing service in the Spanish Civil War.

| | |
|---|---|
| Country of origin: | Spain |
| Calibre: | 7.65mm |
| Length: | 146mm (5.75in) |
| Weight: | 0.57kg (1.25lb) |
| Barrel: | 81mm (3.2in) |
| Feed/magazine capacity: | 7-round detachable box magazine |
| Operation: | Blowback |
| Muzzle velocity: | 229mps (750fps) |
| Effective range: | 30m (98ft) |

# Astra Model 400

The Astra Model 400, also known as the Astra Model 21, was based on an earlier Spanish weapon, the 9mm Campo-Giro Model 1913, and retains the Campo-Giro's distinctive 'air pistol' shape. The Astra entered production in the early 1920s and went on to enjoy a long service life into the 1950s with the Spanish and French armies, and with commercial interests worldwide. In terms of action, there is little exceptional about the Astra – it is a conventional blowback weapon. Its distinction comes from its ability to chamber and fire most 9mm rounds available on the market – even the .38 Colt Auto bullet. Rare versions were also made for 7.65mm ACP and 7.63mm Mauser ammunition. This versatility means that Astra Model 400s still crop up today, although worn pistols lose their ability to fire many modern 9mm rounds safely.

| | |
|---|---|
| **Country of origin:** | Spain |
| **Calibre:** | 9 x 23mm Largo; 7.65mm ACP; 7.63mm Mauser |
| **Length:** | 235mm (9.25in) |
| **Weight:** | 1.08kg (2.38lb) |
| **Barrel:** | 140mm (5.5in), 6 grooves, rh |
| **Feed/magazine capacity:** | 8-round detachable box magazine |
| **Operation:** | Blowback |
| **Muzzle velocity:** | 345mps (1132fps) |
| **Effective range:** | 40m (131ft) |

# Astra Falcon

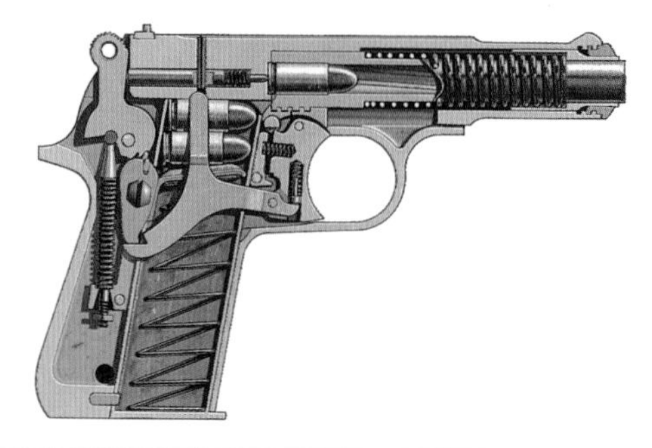

The Astra 'water pistol' design carried the company well through the first half of the 20th century. The Astra Falcon was the last pistol in that line. It appeared in the 1950s when Spain was still under Franco's rule, and is still in service today, owing to its solid build and its sound concept as an automatic handgun. In operation it is indistinguishable from the Astra 400 – indeed the Falcon is essentially the 400 design reduced considerably in length and weight to make it a handgun more suited to post war police use. Whereas the 400 model was some 235mm (9.25in) long, the Falcon was scaled down to 164mm (6.4in) and its weight was reduced to a more comfortable 0.64kg (1.4lb). Unfortunately, the Astra Falcon is a lightweight gun in more than one respect, as its 9mm Short (.380 Auto) cartridge is generally a poor performer, and not well suited to the police role.

| | |
|---|---|
| Country of Origin: | Spain |
| Calibre: | 9mm Short (.380 Auto) |
| Length: | 164mm (6.4in) |
| Weight: | 0.646kg (1.4lb) |
| Barrel: | 98.5mm (3.8in), 4 grooves, rh |
| Feed/Magazine capacity: | 7-round detachable box magazine |
| Operation: | Blowback |
| Muzzle Velocity: | c.300mps (984fps) |
| Effective Range: | 30m (98ft) |

# Star 30M

The Star 30M is an excellent pistol which first entered production in 1990 and is still going strong today in various commercial and military hands. On first glance, the 30M actually looks like a standard Browning imitation gun, but further examination shows that the weapon's slide is actually running inside the gun's frame instead of outside. This configuration gives the gun great stability when it is being fired, and therefore increased accuracy. The resultant performance and the double-action trigger system enabled the 30M to take the recent contract for a Spanish Army service pistol and it is likely that this excellent design will open up many export markets for Astra in the future. In addition to the all-steel 30M, there is a light alloy version available on the market known as the 30PK, which has a lighter overall weight.

| | |
|---|---|
| Country of origin: | Spain |
| Calibre: | 9mm Parabellum |
| Length: | 205mm (8.07in) |
| Weight: | 1.14kg (2.51lb) |
| Barrel: | 119mm (4.6in), 6 grooves, rh |
| Feed/Magazine capacity: | 15-round detachable box magazine |
| Operation: | Blowback |
| Muzzle Velocity: | 380mps (1250fps) |
| Effective Range: | 40m (131ft) |

# SIG P-210

SIG's reputation for excellence in pistol design and manufacture was taken to new levels with the production of the P-210 pistol by the Swiss manufacturer shortly after World War II. Looked at from every angle, the P-210 is a superb gun which, like the P-226 in the US Army pistol trials in the 1980s, fell short of its potential military sales through its high price. A studied glance at the pistol confirms one of its most distinctive features – the slide is situated inside the frame and runs on rails. This means that the pistol hold its accuracy extremely well and its short recoil mechanism is very robust and dependable. The SIG P-210 entered military service in Denmark and police service in West Germany, and was sold commercially across the world. The P-210 handgun is still in production today in various target and operational configurations.

| | |
|---|---|
| Country of origin: | Switzerland |
| Calibre: | 9 x 19mm Parabellum |
| Length: | 215mm (8.46in) |
| Weight: | 0.9kg (1.98lb) |
| Barrel: | 120mm (4.7in), 6 grooves, rh |
| Feed/magazine capacity: | 8-round detachable box magazine |
| Operation: | Short recoil |
| Muzzle velocity: | 335mps (1100fps) |
| Effective range: | 40m (131ft) |

# SIG-Sauer P-225

The SIG-Sauer P-225 shares the P-220's superlative reputation for quality, accuracy and reliability. It is actually a slightly compressed version of the P-220, losing nearly 2cm (0.8in) in length, and was developed to equip the West German police in the 1970s. That particular client needed a gun with failure-proof safety systems, so SIG-Sauer introduced a new automatic locking system for the firing pin which meant that, even if the hammer were accidentally knocked forwards, the gun would not fire. The West German police service accepted the gun, as did the Swiss police, various US security agencies and, it is rumoured, certain US special forces units. Magazine capacity in the P-225 remained at eight 9mm rounds; subsequently, the P-226 variant took that capacity up to 15 rounds to make it more suitable for military use.

| | |
|---|---|
| Country of origin: | Switzerland (SIG); Germany (J. P. Sauer & Sohn) |
| Calibre: | 9mm Parabellum |
| Length: | 180mm (7.08in) |
| Weight: | 0.74kg (1.63lb) |
| Barrel: | 98mm (3.85in), 6 grooves, rh |
| Feed/magazine capacity: | 8-round detachable box magazine |
| Operation: | Short recoil |
| Muzzle velocity: | 340mps (1115fps) |
| Effective range: | 40m (131ft) |

# SIG-Sauer P-226

The SIG-Sauer P-226 very nearly became the US Army's standard handgun following the acrimonious pistol trials in the mid-1980s. It was only just beaten in the competition by the Beretta Model 92 on the issue of price alone, and its superb quality has seen the design taken into use amongst many US police and security agencies such as the FBI (FBI agents use the gun with subsonic loads more appropriate to their police work). It relies heavily on parts from the earlier SIG P-220 and P-225 designs, yet it differs in several respects relevant to its original military intention. The magazine is expanded to take a full fifteen 9mm rounds, seven rounds more than the P-225. Furthermore, the gun is designed for ambidextrous use, with a magazine release on both sides of the gun just behind the trigger.

| | |
|---|---|
| Country of origin: | Switzerland |
| Calibre: | 9 x 19mm Parabellum |
| Length: | 196mm (7.72in) |
| Weight: | 0.75kg (1.65lb) |
| Barrel: | 112mm (4.41in), 6 grooves, rh |
| Feed/magazine capacity: | 15-round detachable box magazine |
| Operation: | Short recoil |
| Muzzle velocity: | 350mps (1148fps) |
| Effective range: | 40m (131ft) |

# .32 Savage Model 1907

The .32 Savage Model 1907 was a fairly well-rounded gun, although it failed to be adopted by the US forces after it was overshadowed by the legendary Colt M1911 in military trials. After these trials, its manufacturers, the Savage Arms Corporation of Chicopee Falls, Massachusetts, needed a customer; in 1914, they found it in the Portugese armed forces, which had been severed from their German gun supplies. The Model 1907, and its successors in 1908 and 1915, were of a delayed-blowback action, the retardation created by the barrel turning through lugs before the slide could move backwards. The Savage was let down by it being possible for the firing pin actually to touch a round in the chamber – a hard knock could thus set it off – which necessitated the annoying unloading of the gun between firing.

| | |
|---|---|
| Country of origin: | USA |
| Calibre: | .32 ACP |
| Length: | 165mm (6.5in) |
| Weight: | 0.63kg (1.37lb) |
| Barrel: | 95mm (3.75in), 6 grooves, rh |
| Feed/magazine capacity: | 10-round detachable box magazine |
| Operation: | Delayed blowback |
| Muzzle velocity: | 244mps (800fps) |
| Effective range: | 30m (98ft) |

# Colt Police Positive

The Police Positive range of Colt revolvers began life in 1905, a development from the earlier range of Pocket Positive revolvers which had achieved popularity in the United States. Throughout its production life the gun was issued in a wide variety of different calibres to suit different roles, and the barrel lengths also fluctuated from a target version (specifications below) to snub nose versions. The virtue of the Police Positive range was primarily that of most Colt handguns – they worked and were capable of taking a man down (early models suffered from a lack of power which was amended by the introduction of a new S&W .38 round). They were also light and were balanced in firing. Police Positive guns or their derivations remained in production into the 1960s, and their design heritage is still influential today.

| | |
|---|---|
| Country of origin: | United States |
| Calibre: | 0.22in |
| Length: | 260mm (10.25in) |
| Weight: | 0.68kg (1.5lb) |
| Barrel: | 152mm (6in) |
| Feed/magazine capacity: | 6-round cylinder |
| Operation: | Revolver |
| Muzzle velocity: | 354mps (1161fps) |
| Effective range: | 30m (98ft) |

# Colt M1911

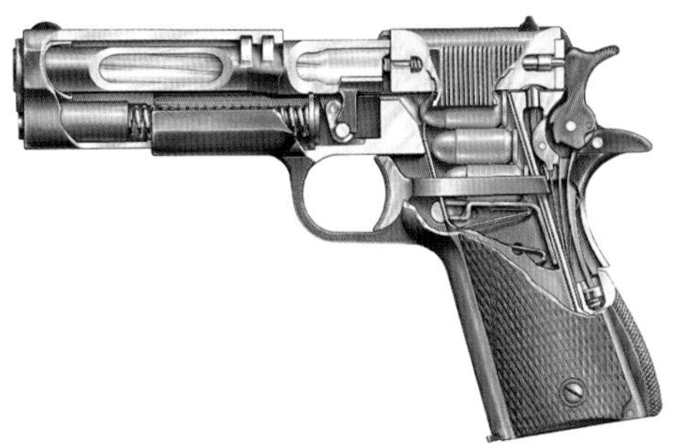

Possibly the most famous and successful handgun of all time, the Colt M1911 and its variants stayed in service with US forces from 1911 to 1990, an incredible longevity. The Colt M1911 was born after a .38 automatic, designed by Browning, proved an ineffective man-stopper during service in the Philippines. Browning and Colt completely redesigned the gun in a powerful .45 calibre, and the US Army accepted it into service as 'U.S Pistol, Automatic, Caliber .45, Model 1911'. The M1911 and the M1911A1 (in production from 1926) achieved awesome success through their simplicity, reliability and power. The swinging-link short-recoil operation was rugged and the .45 round offered conclusive firepower. By the 1980s, the M1911A1 had become dated with its small magazine capacity and its difficulty in handling, and it was finally (and controversially) replaced by the Beretta 92.

| | |
|---|---|
| Country of origin: | United States |
| Calibre: | .45 ACP |
| Length: | 216mm (8.5in) |
| Weight: | 1.13kg (2.49lb) |
| Barrel: | 127mm (5in), 6 grooves, rh |
| Feed/magazine capacity: | 7-round detachable box magazine |
| Operation: | Short recoil |
| Muzzle velocity: | 253mps (830fps) |
| Effective range: | 35m (114ft) |

# Colt New Service Revolver M1917

After a period of using service revolvers of .38 calibre, the US Army turned once again to the potent .45 calibre round because of its combat experiences in the Philippines around the end of the 19th century. Finding that .38 rounds could often go straight through an assailant, rather than put him down, Colt brought out the .45 New Service Revolver (M1909) in 1909, which had greater stopping power. This was the last US service revolver to be adopted before the Colt M1911 automatic pistol took over the role, but the M1909 was improved in 1917 for wartime use as the M1917. The M1917 fired the rimless .45 ACP cartridge, as opposed to the rimmed .45 Colt of the M1909 (which was longer, but generated considerably less muzzle velocity), and loading such rounds into the gun could be speeded up by using three-round clips.

| | |
|---|---|
| Country of origin: | USA |
| Calibre: | .45in ACP |
| Length: | 273mm (10.75in) |
| Weight: | 1.02kg (2.25lb) |
| Barrel: | 140mm (5.5in), 6 grooves, lh |
| Cylinder capacity: | 6 rounds |
| Operation: | Revolver |
| Muzzle velocity: | 282mps (925fps) |
| Effective range: | 35m (115ft) |

# .45in Colt Army Model 1917

Like the Smith & Wesson Model 1917, the Colt Model 1917 was one of the pistols purchased by the US Army which were converted to take the rimless .45in ACP cartridge and used to plug the military pistol shortages which had developed at the time due to the USA entering World War I. Again, like the Smith & Wesson, the new round necessitated the shortening of the cylinder and loading using three-round clips which enabled the rimless rounds to be ejected after firing. Prior to its conversion, the M1917 had been the Colt New Service Revolver, which began its life in 1897 and ranged across 18 different calibres. When standardised, some 150,000 M1917s were purchased by the US Army and some also crossed the Atlantic to the British Army in .445 calibre. The M1917's service outlasted World War I and continued until the end of the next world war.

| | |
|---|---|
| Country of origin: | USA |
| Calibre: | .45in ACP |
| Length: | 272mm (10.75in) |
| Weight: | 1.14kg (2.51lb) |
| Barrel: | 140mm (5.5in), 6 grooves, lh |
| Cylinder capacity: | 6 rounds |
| Operation: | Revolver |
| Muzzle velocity: | 265mps (870fps) |
| Effective range: | 30m (98ft) |

# Colt Detective Special

The legendary Colt Detective Special was in effect a Police Positive revolver rendered in a shorter 50mm (2in) barrel for the ultimate in portability and concealable convenience. The short barrel meant that the gun could be drawn quickly from a holster. It was also an especially light weapon, weighing in at only 0.6kg (1.3lb). Despite the extremely short barrel length, the gun remained businesslike in terms of power, owing to the .38 Special cartridge it fired. Although production of the Detective Special began in 1926, the gun was still going strong in the 1950s; yet, by then, a plethora of Colt variations were emerging. The standard Detective Special thus became the Model D.1 to distinguish it from other weapons. One feature which can be seen on some Detective Specials is a detachable hammer shroud which could be both self- and factory-fitted.

| | |
|---|---|
| Country of origin: | USA |
| Calibre: | .38 Special |
| Length: | 171mm (6.7in) |
| Weight: | 0.06kg (1.31lb) |
| Barrel: | 54mm (2.13in), 6 grooves, rh |
| Cylinder capacity: | 6 rounds |
| Operation: | Revolver |
| Muzzle velocity: | 213mps (700fps) |
| Effective range: | 30m (98ft) |

# Colt Python

**Since its beginnings in 1955, the Colt Python has become one of the world's finest combat handguns. This reputation comes in the main from its high levels of workmanship and finish, and today it is made in either stainless steel or blued carbon steel. The gun's one drawback (apart from its very high price tag due to its quality of manufacture) is its high weight – around 1.16kg (2.56lb) – although this is an essential part of the gun's ability to fire the potent .357 Magnum round for which it is principally chambered. (A small batch of Pythons were made around the .38 Smith &Wesson cartridge). This weight does, however, make the gun incredibly durable and, if the user is strong enough to handle it, then there are few guns that will give better, more consistent performance, and the Magnum round gives the Python a great deal of stopping power.**

| | |
|---|---|
| Country of origin: | USA |
| Calibre: | .357 Magnum |
| Length: | 235mm (9.25in) to 343mm (13.5in) |
| Weight: | 1.08kg (2.37lb) to 1.2kg (2.62lb) |
| Barrel: | 102mm (4in); 204mm (8in) |
| Cylinder capacity: | 6 rounds |
| Operation: | Revolver |
| Muzzle velocity: | 455mps (1500fps) |
| Effective range: | 50m (164ft) |

# Ruger Security Six

The Ruger Security Six is a very powerful weapon, a power which it derives from its potent .357 Magnum round with its awesome stopping ability. The gun was produced by Ruger from 1968, but the design actually came from an earlier single-action revolver that was developed in the 1950s. Ruger's intention behind the Security Six was to create a quality handgun for use as a standard police weapon. The choice of round gave police officers a more versatile response than lighter calibres such as the .38, which had reduced penetrative power when fired against vehicles or other light constructions such as buildings. Sturm, Ruger & Co. has continued to make .357 weapons for the police market, and today models such as the GP 100 are continuing to build Ruger's reputation for accuracy and controlled power in its handguns.

| | |
|---|---|
| Country of origin: | United States |
| Calibre: | .357 Magnum |
| Length: | 235mm (9.25in) |
| Weight: | 0.95kg (2.09lb) |
| Barrel: | 102mm (4in), 6 grooves, rh |
| Feed/magazine capacity: | 6-round cylinder |
| Operation: | Revolver |
| Muzzle velocity: | c.400mps (1312fps) |
| Effective range: | 40m (131ft) + |

# Ruger Redhawk

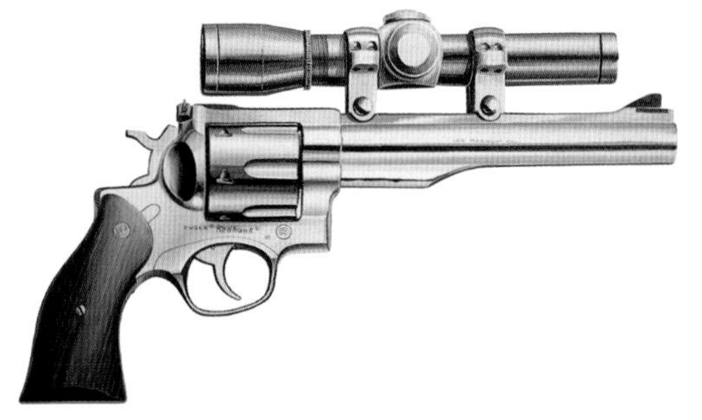

Sturm, Ruger & Company's first double-action revolvers were the Security Six, Police Six and Speed Six, which emerged in the 1970s, but are now discontinued; the Redhawk was a continuation of the theme of a powerful .44 Magnum handgun for use primarily by hunters in the United States rather than the military. The potential force within the gun is indicated by the Redhawk model's integral telescopic sight mounts being cut into the length of the barrel. The sight coupled with the powerful round enable the Ruger to be turned into a hunting weapon with an effective range of around 60m (197ft). To take the .44 round's power, the barrel has an increased wall thickness just in front of the chamber. The Redhawk's combination of compactness and sheer brute power has made it a popular weapon for commercial users throughout the world.

| | |
|---|---|
| Country of origin: | USA |
| Calibre: | .44 Magnum |
| Length: | 165mm (6.5in) |
| Weight: | 1.5kg (3.37lb) |
| Barrel: | 190mm (7.48in) |
| Cylinder capacity: | 6 rounds |
| Operation: | Revolver |
| Muzzle velocity: | 450mps (1475fps) |
| Effective range: | 60m (197ft) |

# Smith & Wesson Double-Action

After some four years of research, during the 1880s, Smith & Wesson advanced its range of gun designs with a new series of double-action revolvers (i.e. revolvers that can be fired through both a single pull on the trigger which takes the hammer through its full cycle, or by manually cocking the hamer then releasing it through a shortened trigger pull). Using the break-open system of ejection and reloading with the barrel section hinged in front of the cylinder, they were popular and reliable guns that fired Smith &Wesson's .38 bullet to an effective range of around 20m (66ft). The double-action range was proficient enough to endure beyond the 19th century to the 1940s, during which time side-opening cylinder weapons became more prevalent in weapon design and eventually superseded break-open guns.

| | |
|---|---|
| Country of origin: | USA |
| Calibre: | .38 S&W |
| Length: | 190mm (7.5in) |
| Weight: | 0.51kg (1.12lb) |
| Barrel: | 83mm (3.27in), 5 grooves, rh |
| Feed/magazine capacity: | 6-round cylinder |
| Operation: | Revolver |
| Muzzle velocity: | 190mps (625fps) |
| Effective range: | 20m (66ft) |

# Smith & Wesson 'Hand Ejector' Model 1917

The Model 1917 was part of a US action to remedy a shortage of front-line pistols among US Army troops on the Western Front during World War I. As the standard service revolver at the time (Colt M1911 automatic) was .45 ACP calibre, the US forces purchased more than 150,000 Smith & Wesson revolvers and converted them to fire that round. This weapon thus become known as the M1917. Being a rimless round, the .45 ACP had to be loaded into the revolver in special three-round clips, otherwise the ejectors had no edge on which to gain purchase for cartridge removal. This situation was remedied on the commercial market in the 1920s by the Peters Cartridge Company, which developed a .45 rimmed round called the '.45 Auto Rim'. The M1917 was a good, solid revolver which gave service well beyond the end of the war in 1918.

| | |
|---|---|
| Country of origin: | USA |
| Calibre: | .45in ACP |
| Length: | 274mm (10.80in) |
| Weight: | 1.02kg (2.25lb) |
| Barrel: | 140mm (5.5in), 6 grooves, lh |
| Cylinder capacity: | 6 rounds |
| Operation: | Revolver |
| Muzzle velocity: | 265mps (870fps) |
| Effective range: | 40m (132ft) |

# Smith & Wesson Mk 22 Model 0 'Hush Puppy'

The infamous 'Hush Puppy' was born out of the demand by US Navy SEAL teams in Vietnam for an effective silenced pistol with which to dispose of Vietcong soldiers and sentry animals during covert missions. One answer was a suppressed Walther P38; the other was to extend the barrel of Smith & Wesson's Model 39 by 12.7cm (5in), thread the end and fit a silencer. With a slide lock and adjustable sights, and firing the subsonic Mark 144 Model 0 ammunition, this latter weapon became designated as the Mk22 Model 0 and was introduced into service in 1967. Known mainly by its nickname, the 'Hush Puppy' served the elite units well. Its muzzle velocity of 294.13mps (965fps) gave it enough power and range to be used at reasonable distances (accuracy allowing) and it stayed in military use until the 1980s.

| | |
|---|---|
| Country of origin: | United States |
| Calibre: | 9mm Parabellum |
| Length: | 323mm (12.75in) |
| Weight: | 0.96kg (2.1lb) |
| Barrel: | 101mm (3.9in) 6 grooves, rh |
| Feed/magazine capacity: | 8-round detachable box magazine |
| Operation: | Blowback |
| Cyclic rate of fire: | Semi-automatic |
| Muzzle velocity: | 274mps (900fps) |
| Effective range: | 30m (98ft) |

# Smith & Wesson 459

**The Model 459 was one of Smith & Wesson's contributions to the US Army pistol trials in the 1980s, an event which turned into an acrimonious legal action after S&W were quickly rejected in favour of European weapons manufacturers (Beretta were the final winners with their 92SB model). Smith & Wesson's original entry was the 469, but after this fell by the wayside in the first contest, the 459 was substituted and submitted into the second. The 459's heritage lies in the Model 39, produced between 1954 and 1980. This was a 9mm gun which was used by the US Navy and Special Forces, but which became limited through only having an 8-round magazine. To rectify this a 14-round version was produced, the Model 59, and an improved version of this became the 459. Both the Model 59 and 459 were designed specifically for military use.**

| | |
|---|---|
| Country of origin: | United States |
| Calibre: | 9mm Parabellum |
| Length: | 175mm (6.89in) |
| Weight: | 0.73kg (1.6lb) |
| Barrel: | 89mm (3.5in), 6 grooves, rh |
| Feed/magazine capacity: | 14-round detachable box magazine |
| Operation: | Blowback |
| Muzzle velocity: | 395mps (1295fps) |
| Effective range: | 40m (131ft) |

# Smith & Wesson Model 29 .44 Magnum

**The power of the .44 Magnum handgun gained particular notoriety in the hands of Clint Eastwood in the 1970s, through his role in the popular 'Dirty Harry' movies (Smith & Wesson actually launched the Model 29 back in 1955). The Model 29 came with a choice of three barrel lengths, ranging from the shortest – a 102mm (4in) version – to a striking 203mm (8.6in) version (which appeared in the films, and to which the specifications below refer). Production of the Model 29 by Smith & Wesson has continued to this day in several different versions, now mostly designated as the 629 series. Most of the Model 29/629 series are distinguished by finish (the 629 first came out in stainless steel in 1979), barrel length or the positioning of the ejector rod, which is either shrouded or set into a recess underneath the barrel.**

| | |
|---|---|
| Country of origin: | USA |
| Calibre: | .44in Magnum |
| Length: | 353mm (13.89in) |
| Weight: | 1.45kg (3.19lb) |
| Barrel: | 203mm (8in), 6 grooves, rh; also 153mm (6in) or 102mm (4in) |
| Cylinder capacity: | 6 rounds |
| Operation: | Revolver |
| Muzzle velocity: | 450mps (1476fps) |
| Effective range: | 50m (164ft) |

# Smith & Wesson 1006

During the 1980s, Smith & Wesson introduced a Third Generation series of pistols in various calibres ranging from 9mm to .45 ACP and including, in the case of the 1006, a 10mm Auto weapon. This was not the first 10mm development in the history of handguns – the US company Dornaus & Dixon produced one, the Bren 10, in the early 1980s, but this ceased production when the company went into receivership in 1985. By this time, however, the FBI were interested in the potential of the 10mm gun and so Smith & Wesson produced the 1006 from 1990. The calibre makes the gun very powerful indeed, but good-quality Novak sights and combat grips make it controllable in trained hands. The numbering system is part of Smith & Wesson's Third Generation notation, the first two digits indicating the calibre of the weapon.

| | |
|---|---|
| Country of origin: | USA |
| Calibre: | 10mm Auto |
| Length: | 216mm (9in) |
| Weight: | 1.07kg (2.37lb) |
| Barrel: | 127mm (5in), 6 grooves, rh |
| Operation: | Short recoil |
| Feed/magazine capacity: | 9-round detachable box magazine |
| Muzzle velocity: | 335mps (1100fps) |
| Effective range: | 30m (98ft) |

# Liberator

The word 'cheap' does not begin to describe the Liberator, also known as the .45 OSS after its primary buyer, the Office of Strategic Services. The intention behind this weapon was to mass produce a single-shot weapon which could be used by resistance forces during the last stages of World War II. The gun was made out of the simplest possible steel pressings and stampings, ejection had to be accomplished using a short stick, the breech block was hand operated, and it had no rifling. Yet this crudity meant production was awesome – one million in three months alone. The gun was supplied with 10 rounds of ammunition and a set of univerally understandable cartoon-style instructions. Having almost no accuracy, it was intended primarily as an assassination tool. How many accomplished this task we will never know.

| | |
|---|---|
| Country of origin: | USA |
| Calibre: | .45 ACP |
| Length: | 141mm (5.55in) |
| Weight: | 0.45kg (1lb) |
| Barrel: | 101mm (3.97in), smooth bore |
| Feed/magazine capacity: | 1 round inserted directly into chamber |
| Operation: | Single shot |
| Muzzle velocity: | 250mps (820fps) |
| Effective range: | 10m (32.8ft) |

# 13mm MBA Gyrojet

It is a contentious notion as to whether the Gyrojet is in fact a pistol at all. It is, in effect, a hand-held rocket launcher, the brainchild of two US inventors in the 1960s: Robert Mainhardt and Art Biehl. Instead of firing standard unitary cartridge rounds, the Gyrojet was designed to take 13 x 38mm projectiles which consisted of a solid or explosive-filled head fixed into a tubular body that contained a propellant and which had a baseplate containing a percussion cap and four thruster jet apertures. When fired, the percussion cap ignited the propellant, which in turn blasted the round out of the barrel. There being no cartridge to extract, the next round simply popped into the chamber ready for firing. Despite the fact that the jets were angled to give gyroscopic spin to the round in flight, the Gyrojet was not accurate and it never went beyond conceptual production.

| | |
|---|---|
| Country of origin: | USA |
| Calibre: | 13mm rocket projectiles |
| Length: | 234mm (9.2in) |
| Weight: | 0.98kg (2.1lb) |
| Barrel: | 127mm (5in) |
| Feed/magazine capacity: | 6-round detachable box magazine |
| Operation: | Short recoil |
| Muzzle velocity: | 274mps (900fps) |
| Effective range: | 50m (164ft) |

# Calico M950

One of the most striking modern pistols, the Calico M950 strays towards the submachine gun category, even though it will only fire in a semi-automatic mode. As is evident from the illustration, the grip for the weapon is two-handed and, with the magazine, the length of the M950 extends to 365mm (14.3in). Undoubtedly, it is the feed system of the Calico which is its most distinctive feature. The 50- or 100-round helical magazines are fitted to the top of the weapon running lengthwise down the receiver (50-round magazines are the more standard). Weighing about 1.8kg (3.97lb) when loaded (1kg (2.2lb) when empty), the M950 fires standard 9mm Parabellum rounds to an effective range of about 60m (197ft) on account of its longer barrel. An interesting weapon, the Calico's military markets are yet to be decided.

| | |
|---|---|
| Country of origin: | USA |
| Calibre: | 9mm Parabellum |
| Length: | 365mm (14.3in) |
| Weight: | 1kg (2.2lb) |
| Barrel: | 152mm (6in), 6 grooves, rh |
| Feed/magazine capacity: | 50- or 100-round detachable helical magazine |
| Operation: | Delayed blowback |
| Muzzle velocity: | 393mps (1290fps) |
| Effective range: | 60m (197ft) |

# Nagant M1895

Although the Nagant M1895 was produced, as its name reveals, in the late 19th century while Russia still had a Tsar, it continued in service until the 1950s with the Soviet Union and was a remarkable weapon. Its most significant contribution to weapon history is that it used a means of completely eradicating the fractional loss of gas pressure between the cylinder and barrel common to the vast majority of revolvers. It achieved this by pushing the cylinder forwards onto the tapered barrel end when the hammer was cocked, which in turn allowed the long Nagant cartridge, which held the bullet entirely in its case length, to enter directly into the barrel and produce complete obduration on firing. The Nagant's long service record is a testimony to the success of this procedure and it was a prized weapon for many years.

| | |
|---|---|
| Country of origin: | USSR/Russia |
| Calibre: | 7.62 x 38R Russian Revolver |
| Length: | 203mm (8in) |
| Weight: | 0.89kg (1.96lb) |
| Barrel: | 114mm (4.5in), 6 grooves, rh |
| Cylinder capacity: | 7 rounds |
| Operation: | Single- or double-action revolver |
| Muzzle velocity: | 305mps (1000fps) |
| Effective range: | 30m (98ft) |

# Tula-Tokarev TT-33

**The Tula-Tokarev 33 (TT-33) began in the late 1920s as the TT-30, an automatic pistol designed by Feodor Tokarev and based on the US M1911's swinging-link locking system. Tokarev simplified its components for the rigours of Soviet service, which involved machining the magazine feed lips onto the receiver itself, making the hammer and lock system in a detachable module at the back (more convenient for both manufacture and repair) and taking away the safety devices. The TT-33 became the Tokarev's dominant form from 1933. This differed from the TT-30 by having locking logs sited all around the barrel instead of just on top, something that speeded up production as barrel and locking lugs could be done at the same time. The TT-33 is still manufactured under licence today, in countries such as Poland and North Korea.**

| | |
|---|---|
| **Country of origin:** | USSR/Russia |
| **Calibre:** | 7.62mm Soviet |
| **Length:** | 193mm (7.68in) |
| **Weight:** | 0.83kg (1.83lb) |
| **Barrel:** | 116mm (4.57in), 4 grooves, rh |
| **Feed/magazine capacity:** | 8-round detachable box |
| **Operation:** | Short recoil |
| **Muzzle velocity:** | 415mps (1362fps) |
| **Effective range:** | 30m (98ft) |

# Pistole Makarov

**The standard Soviet side arm from the late 1950s, the Pistole Makarov is recognisably based on the excellent Walther PP. Features transferred from the PP into the Makarov – although without as much elegance of form and function – are the double-action trigger pull and the locking/unlocking catch for the hammer. Where it does diverge, however, is in the calibre. The Makarov uses a 9 x 18mm cartridge that falls somewhere between the 9mm Parabellum and the 9mm Short, thus making the gun incompatible with much of the Western world's ammunition, but still enabling it to operate safely on the blowback system. Despite being somewhat bulkier than the Walther model, the Makarov is a solid performer regardless of environment and it spread throughout the communist world, particularly to China (the Type 59) and the former East Germany (Pistole M).**

| | |
|---|---|
| Country of origin: | USSR/Russia |
| Calibre: | 9 x 18mm Soviet |
| Length: | 160mm (6.3in) |
| Weight: | 0.663kg (1.46lb) |
| Barrel: | 91mm (3.5in), 4 grooves, rh |
| Feed/magazine capacity: | 8-round detachable box |
| Operation: | Blowback |
| Muzzle velocity: | 315mps (1033fps) |
| Effective range: | 40m (131ft) |

# Stechkin

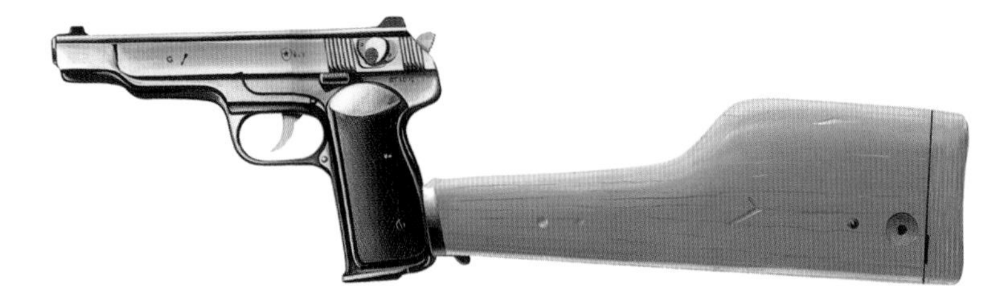

**The Stechkin was one of the Soviet Union's less successful postwar experiments in small arms design. The intention was to create a compact machine pistol with a fairly high rate of fire; this was pursued by producing a gun roundly based on the Walther PP, but with full-auto capability. It worked by blowback, using the mid-powered 9mm Soviet cartridge and could be fitted with a shoulder stock to give the user greater control during automatic fire, which had a cyclical rate of 850rpm. The Stechkin, however, was effectively too small for a machine pistol and too big for a pistol. It was fairly uncontrollable on full automatic and too bulky for conventional handling as a self-loading weapon, thereby satisfying nobody. Nonetheless, it stayed in production between 1951 and 1975, before finally being withdrawn from troop use.**

| | |
|---|---|
| **Country of origin:** | USSR/Russia |
| **Calibre:** | 9mm Makarov |
| **Length:** | 225mm (8.86in) |
| **Weight:** | 1.03kg (2.27lb) |
| **Barrel:** | 127mm (5in), 4 grooves, rh |
| **Feed/magazine capacity:** | 20-round detachable box magazine |
| **Operation:** | Blowback |
| **Muzzle velocity:** | 340mps (1115fps) |
| **Effective range:** | 30m (98ft) |

# PSM

The PSM (Pistolet Samozaryadniy Malogabaritniy, meaning 'Pistol, Self-Loading, Small') appears to have been designed with concealability in mind, as it has a narrow frame and smooth surfaces, and mainly (to our knowledge in the West) entered service with the Soviet police forces and special military units. It has since gone on to be a part of criminal kit throughout central Europe and modern-day Russia. The PSM is a double-action blowback weapon working on the rather low-powered 5.45mm Soviet Pistol cartridge. In overall appearance and much of its action, it resembles the Walther PP, but it does not appear to have the sophistication and quality inherent in the German weapon. Nevertheless, the PSM will probably have a long service life, if only in the hands of criminals, who can conveniently fit it into a pocket.

| | |
|---|---|
| Country of origin: | USSR/Russia |
| Calibre: | 5.45mm Soviet Pistol |
| Length: | 160mm (6.3in) |
| Weight: | 0.46kg (1.01lb) |
| Barrel: | 85mm (3.35in), 6 grooves, rh |
| Feed/magazine capacity: | 8-round detachable box magazine |
| Operation: | Blowback |
| Muzzle velocity: | 315mps (1033fps) |
| Effective range: | 40m (131ft) |

# Austen MK1

The Austen was the Australian attempt to produce a wartime submachine gun that was of better quality and reliability than the British Stens which the Australian forces were receiving from 1941. Their answer was a fusion of the Sten and the German MP40. Thus, while the barrel, receiver and trigger belonged to the Sten, the mainspring, bolt and stock are from the MP40. The combination was a successful one and the Austen proved to be a thoroughly competent weapon in action, with a greater durability than the Sten. Thus the Austen and Owen submachine guns were able to equip the Australian forces. Yet, though some 20,000 Austens were made, it was the Owen that dominated, mainly because it had a reliability that the Austen could not match.

| | |
|---|---|
| Country of origin: | Australia |
| Calibre: | 9mm Parabellum |
| Length: | 845mm (33.25in) stock extended; 552mm (21.75in) stock folded |
| Weight: | 3.98kg (8.75lb) |
| Barrel: | 196mm (7.75in), 6 grooves, rh |
| Feed/magazine capacity: | 28-round detachable box magazine |
| Operation: | Blowback |
| Cyclic rate of fire: | 500rpm |
| Muzzle velocity: | 380mps (1246fps) |
| Effective range: | 50m (164ft) |

# Owen

Australia is one of the few nations to have experimented with top-mounted magazines on submachine guns – in the Owen, the experiment paid off. The Owen was adopted as a last-ditch measure in 1940, when the United Kingdom was unable to supply Australia with Sten guns. Despite its appearance, the Owen was comfortable to use and the vertical feed system worked efficiently and reliably, and it soon became popular with Australian frontline troops. The basic Owen came in two variants: early models had a solid frame and featured barrel-cooling fins; the later, more numerous model had a skeletal stock and the cooling fins were removed. Although heavy and featuring necessarily offset sights, the Owen had no serious defects and its use after the war continued into the 1960s.

| | |
|---|---|
| **Country of origin:** | Australia |
| **Calibre:** | 9mm Parabellum |
| **Length:** | 813mm (32in) |
| **Weight:** | 4.21kg (9.28lb) |
| **Barrel:** | 247mm (9.75in), 7 grooves, rh |
| **Feed/magazine capacity:** | 33-round detachable box magazine |
| **Operation:** | Blowback |
| **Cyclic rate of fire:** | 700rpm |
| **Muzzle velocity:** | 380mps (1247fps) |
| **Effective range:** | 70m (230ft) |

# Steyr MP69

The Steyr Mpi69 is a well-conceived, solid submachine gun that stayed in production between 1969 and 1980. Compact, and with a 550rpm rate of fire, the MPi69 is a blowback weapon with the unusual feature of fire selection being performed by the trigger pull – pull it half-way back for single shots, and all the way back for full automatic fire. The MPi69 needs to be distinguished from a later model, the MPi81. The MPi69 has the sling attached to the cocking handle, whereas the MPi81 has a conventional sling attachment and a greater rate of fire (700rpm). The Mpi69's excellent strong construction – the barrel is cold hammered for endurance – has made it a popular export weapon, and it still serves with the Austrian Army in various capacities.

| | |
|---|---|
| Country of origin: | Austria |
| Calibre: | 9 x 19mm Parabellum |
| Length: | 670mm (26.38in) stock extended; 465mm (18.3in) stock folded |
| Weight: | 3.13kg (6.9lb) |
| Barrel: | 260mm (10.23in), 6 grooves, rh |
| Feed/magazine capacity: | 25- or 32-round detachable box magazine |
| Operation: | Blowback |
| Cyclic rate of fire: | 550rpm |
| Muzzle velocity: | 380mps (1247fps) |
| Effective range: | 200m (656ft) |

# Vigneron

There is little exceptional about the Belgian Vigneron submachine gun, though its appearance is slightly unusual on account of the long ribbed barrel, which is fitted with a compensator and muzzle brake. It was a child of the 1950s – constructed from a simple and economical metal-stamping method, and made initially for the Belgian Army, seeing active service in the Belgian Congo in the 1960s and subsequently throughout Africa in the hands of various armies once the Belgians had departed. The Vigneron features a pistol grip with an integral safety lever. An unusual feature is its ability to deliver single shots with only a partial pull of the trigger. Its wire stock can also be adjusted to suit the personal dimensions of the user.

| | |
|---|---|
| Country of origin: | Belgium |
| Calibre: | 9 x 19mm Parabellum |
| Length: | 890mm (35in) stock extended; 705mm (27.75in) stock folded |
| Weight: | 3.29kg (7.25lb) |
| Barrel: | 305mm (12in), 6 grooves, rh |
| Feed/magazine capacity: | 32-round detachable box magazine |
| Operation: | Blowback |
| Cyclic rate of fire: | 550rpm |
| Muzzle velocity: | 365mps (1200fps) |
| Effective range: | 200m (656ft) plus |

# FN P90 Personal Defence Weapon (PDW)

The P90 represents perhaps the future of the submachine gun. Its amorphous design houses unique features throughout. The clear plastic magazine lies across the top of the weapon, with the bullets at a right angle to the barrel and passing through a turntable before being loaded into the chamber. An optical sight is balanced by open sights on both sides for left- or right-handed shooting, and all the gun's operating systems are ambidextrous. Only 400mm (15.75in) long, the P-90 has a receiver that acts as a stock and cartridge ejection is down through the hollow pistol grip. Despite these innovations, the mechanism remains simple blowback, but the cartridge is a new high-power 5.7mm FN which gives excellent range and penetration. The future popularity and use of such a weapon are yet to be seen.

| | |
|---|---|
| Country of origin: | Belgium |
| Calibre: | 5.7mm FN |
| Length: | 400mm (15.75in) |
| Weight: | 2.8kg (6.17lb) |
| Barrel: | 263mm (7.75in), 6 grooves, rh |
| Feed/magazine capacity: | 50-round detachable box magazine |
| Operation: | Blowback |
| Cyclic rate of fire: | 800-1000rpm |
| Muzzle velocity: | 850mps (2800fps) |
| Effective range: | 200m (656ft) plus |

# Samopal CZ Model 25

The CZ Model 25 is part of a group of four weapons which began life with the M48A in 1948 (renamed the CZ 23 in 1950). The four weapons are effectively the same except that the CZ 25 has a folding metal stock instead of the wooden stock of the CZ 23, while the CZ 24 and CZ 26 took over from the CZ 23 and CZ 25 in Czech Army service in 1951, these latter guns firing the more powerful 7.62mm Soviet pistol cartridge. Common to all is a distinctive wrapround bolt mechanism: the bolt has a tubular configuration and encloses the rear of the barrel at firing, and the magazine is seated in the pistol-grip itself. Both features contributed to the shortness of the weapon and its general good performance kept it in Czech service until the mid-1960s and, in terrorist hands, to the present day.

| | |
|---|---|
| Country of origin: | Czechoslovakia |
| Calibre: | 9mm Parabellum |
| Length: | 686mm (27in) stock extended; 445mm (17.52in) stock folded |
| Weight: | 3kg (6.75lb) |
| Barrel: | 284mm (11.2in), 6 grooves, rh |
| Feed/magazine capacity: | 24- or 40-round box magazine |
| Operation: | Blowback |
| Cyclic rate of fire: | 600rpm |
| Muzzle velocity: | 395mps (1300fps) |
| Effective range: | 120m (400ft) |

# Samopal 62 'Skorpion'

This incredibly compact weapon was designed to give tank and vehicle crews (who have little storage space) more advanced firepower than that provided by a mere pistol. Only 270mm (11in) long, the Skorpion was capable of putting out 850rpm. More would be possible were it not for a clever restraint on the blowback system in the form of a weight which is driven down into the grip onto a spring while the bolt is held by a catch; the weight is then pushed up by the spring, disengages the catch and the bolt is released. The Model 61 was the original in 1963 and several variations have appeared since then, some chambered for different calibre rounds. The Skorpion is a terrifying gun at close-quarters owing to its spray effect and it has become an easily concealed firearm of many terrorists worldwide.

| | |
|---|---|
| Country of origin: | Belgium |
| Calibre: | 7.65mm Browning (.32 ACP) |
| Length: | 170mm (6.75in) |
| Weight: | 0.62kg (1.37lb) |
| Barrel: | 101mm (4in), 6 grooves, rh |
| Feed/magazine capacity: | 7-round detachable box magazine |
| Operation: | Blowback |
| Muzzle velocity: | 290mps (950fps) |
| Effective range: | 30m (98ft) |

# Madsen M50

The Madsen M50 was actually one of a series of guns made by Dansk Industri Syndikat AS Madsen immediately after WWII. The series was initiated by the disappointing M45, but truly established itself with the M46. This blowback weapon borrowed design and production techniques from the British Sten, US M3 and the Russian PPS guns, the result being a simple, serviceable weapon. Non-standard features, however, included a pressed steel receiver which was hinged and, with the barrel removed, could be opened up to inspect the gun's internal mechanism. The M50 differed from the M46 only marginally – the cocking handle was changed so that it did not have to be removed during stripping, but it was actually one of the most successful guns of the range.

| | |
|---|---|
| Country of origin: | Denmark |
| Calibre: | 9 x 19mm Parabellum |
| Length: | 800mm (31.5in) stock extended; 530mm (20.85in) stock folded |
| Weight: | 3.17kg (6.99lb) |
| Barrel: | 197mm (7.75in), 4 grooves, rh |
| Feed/magazine capacity: | 32-round detachable box magazine |
| Operation: | Blowback |
| Cyclic rate of fire: | 550rpm |
| Muzzle velocity: | 380mps (1274fps) |
| Effective range: | 150m (492ft) plus |

# Suomi Model 31

The Suomi Model 31 was the product of Aimo Johannes Lahti, a designer for the Finnish State Arsenal. His first significant product, the 7.63/7.65mm M26, was not a great commercial success, unlike the M31, its successor. Typically for the 1930s, the M31 was a very well machined blowback-operated weapon which was judiciously made for the increasingly important 9mm Parabellum round. It could be loaded with with either 20- or 50-round box magazines or an influential 71-round drum, which was copied by Shpagin for the PPSh-41. The M31 ended its production in 1944, but stayed in use long after the war – although, in the 1950s, it was converted to use the Carl Gustav's 36-round box magazine. Clients for the M31 reached from Switzerland and Poland to South American countries.

| | |
|---|---|
| Country of origin: | Finland |
| Calibre: | 9 x 19mm Parabellum |
| Length: | 870mm (34.25in) |
| Weight: | 4.87kg (10.74lb) |
| Barrel: | 318mm (12.52in), 6 grooves, rh |
| Feed/magazine capacity: | 30- or 50-round detachable box magazine or 71-round drum |
| Operation: | Blowback |
| Cyclic rate of fire: | 900rpm |
| Muzzle velocity: | 400mps (1310fps) |
| Effective range: | 300m (984ft) |

# Konepistooli M44

Produced as an almost direct copy of the superb Soviet PPS-43, the Konepistooli M44 was made into the Finns' own by their adapting it for 9mm Parabellum ammunition. Despite this, it had little to distinguish it from its inspiration, although its feed system was interchangeable between the 71-round Suomi drum and a 50-round box magazine. It would later (mid-1950s) be modified to accept the 36-round Carl Gustav magazine, and this version was known as the M44/46. Coming from such good stock, the M44 proved a perfectly acceptable weapon. There was little to go wrong with its basic blowback system, although its lack of selector switch (it could only fire full automatic) made it seem increasingly archaic compared to the new international assault rifles and it was taken out of service in the late 1960s.

| | |
|---|---|
| Country of origin: | Finland |
| Calibre: | 9mm Parabellum |
| Length: | 825mm (32.48in) stock extended; 623mm (24.53in) folded |
| Weight: | 2.8kg (6.17lb) |
| Barrel: | 247mm (9.72in), 4 or 6 grooves, rh |
| Feed/magazine capacity: | 71-round drum or 50-round box magazine |
| Operation: | Blowback |
| Cyclic rate of fire: | 650rpm |
| Muzzle velocity: | 395mps (1300fps) |
| Effective range: | 70m (230ft) |

# MAS 38

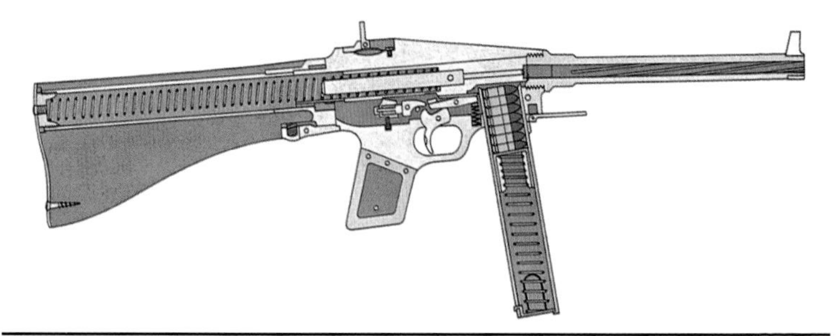

Although the curved line between muzzle and stock gives the MAS 38 an awkward appearance, it was actually a fine weapon. Yet, as its 7.65mm French Long cartridge was only made in France, international sales were not forthcoming. The MAS 38 was a blowback weapon with an especially long bolt travel and a return spring that ran the length of the stock. Recoil was controlled by the low power of the round and the stock's alignment with the barrel, thus resisting muzzle climb. Useful features included a cocking handle that was separate from the bolt when firing and a hinged flap that closed over the magazine aperture when a magazine was withdrawn. The MAS 38 was in production between 1938 and 1949; postwar ammunition standardisation curtailed its future.

| | |
|---|---|
| Country of origin: | France |
| Calibre: | 7.65mm French Long |
| Length: | 734mm (28.9in) |
| Weight: | 2.87kg (6.33lb) |
| Barrel: | 224mm (8.82in), 4 grooves, rh |
| Feed/magazine capacity: | 32-round detachable box magazine |
| Operation: | Blowback |
| Cyclic rate of fire: | 600rpm |
| Muzzle velocity: | 351mps (1152fps) |
| Effective range: | 40m (131ft) |

# MAT 49

The MAT 49 was developed by Manufacture d'Armes de Tulle just after World War II; the company's designers were asked to design a standardised French submachine gun. Made from the most basic processes of stamping and limited machining, the MAT 49 is nonetheless a sound weapon. Its reliability proved itself in the wars in French Indochina and throughout Africa and Southeast Asia in France's former colonies. Indeed, many police and military units still rely on the gun today, although French forces now have the 5.56mm FAMAS rifle as their standard issue. Two interesting features of the MAT 49 are the hinged magazine housing which folds underneath the barrel for storage/carriage, and the way in which the chamber actually wraps around the bolt on firing, rather than the other way round.

| | |
|---|---|
| Country of origin: | France |
| Calibre: | 9mm Parabellum |
| Length: | 720mm (28.35in) stock extended; 460mm (18.11in) stock folded |
| Weight: | 3.5kg (7.72lb) |
| Barrel: | 228mm (8.98in), 4 grooves, rh |
| Feed/magazine capacity: | 20- or 32-round box magazine |
| Operation: | Blowback |
| Cyclic rate of fire: | 600rpm |
| Muzzle velocity: | 390mps (1280fps) |
| Effective range: | 70m (230ft) |

# Bergmann MP18

**The MP18, also known as the Bergmann Muskete, was developed by Hugo Schmeisser in 1916, and heralded the emergence of the blowback submachine gun. It was intended for raids and trench-clearing operations in World War I, a role in which it was briefly – and successfully – tested in 1918. Conditions of the Versailles Treaty kept the MP18 out of German military hands for much of the interwar period, but the weapon was permitted in police hands. It returned to military use after Hitler's rise to power and stayed in production until 1945. The MP18 inspired many new designs with its reliable, straightforward blowback system and 450rpm firepower, and was later developed by Schmeisser into the MP28 (which in turn inspired the British Lanchester).**

| | |
|---|---|
| Country of origin: | Germany |
| Calibre: | 9 x 19mm Parabellum |
| Length: | 815mm (32.09in) |
| Weight: | 4.19kg (9.25lb) |
| Barrel: | 196mm (7.75in), 6 grooves, rh |
| Feed/magazine capacity: | 32-round 'snail' or 20- or 32-round detachable box magazine |
| Operation: | Blowback |
| Cyclic rate of fire: | 450rpm |
| Muzzle velocity: | 395mps (1295fps) |
| Effective range: | 70m (230ft) |

# Erma MPE

The most famous submachine guns to emerge from Erma were the landmark MP38 and MP40, but the MPE was an important step towards those designs. Designed by the talented Heinrich Vollmer, the MPE entered production in 1930, but the rights of manufacture were eventually sold on to Berthold Geipel GmbH in 1934, which took it to larger scale production – in this form it became known as the EMP (Erma Machinenpistole). The distinctive features of the MPE/EMP were its single-shot or full-auto selection facility, wooden foregrip, side-fed magazine and telescoping return-spring casing, the last a subsequent element of the MP38. Customers for the MPE included France, Mexico, Bolivia and Paraguay in the Gran Chaco War (1932–35), and various groups during the Spanish Civil War.

| | |
|---|---|
| Country of origin: | Germany |
| Calibre: | 9mm Parabellum |
| Length: | 902mm (35.5in) |
| Weight: | 4.15kg (9.13lb) |
| Barrel: | 254mm (10in), 6 grooves, rh |
| Feed/magazine capacity: | 20- or 32-round box magazine |
| Operation: | Blowback |
| Cyclic rate of fire: | 500rpm |
| Muzzle velocity: | 395mps (1300fps) |
| Effective range: | 70m (230ft) |

# MP38

**The MP38 gave the German forces exactly what they needed for the blitzkrieg conditions of World War II – a light gun which could provide a high rate of fire and be manufactured in sufficient quantities to meet demand. Its production, which relied on economical metal stampings rather than expensive machining, enabled mass production on a new scale typical of Germany's greater mechanisation of war. Its actual operation was a straightforward blowback and the weapon gained a good reputation for reliability. However, the MP38 was also prone to accidental firing if knocked (as it fired from an open-bolt position) and a modified version, the MP38/40, was produced which used the cocking handle as a locking pin for the breech when not being fired.**

| | |
|---|---|
| **Country of origin:** | Germany |
| **Calibre:** | 9mm Parabellum |
| **Length:** | 832mm (32.75in) stock extended; 630mm (24.75in) stock folded |
| **Weight:** | 4.1kg (9.1lb) |
| **Barrel:** | 247mm (9.75in), 6 grooves, rh |
| **Feed/magazine capacity:** | 32-round box magazine |
| **Operation:** | Blowback |
| **Cyclic rate of fire:** | 500rpm |
| **Muzzle velocity:** | 395mps (1300fps) |
| **Effective range:** | 70m (230ft) |

# MP40

Although the MP38 relied on rationalised methods of manufacture, its replacement, the MP40, took the process further and enabled more than one million to be produced during World War II. The MP40 relied even more on processes of steel pressing, spot welding and sub-assembly than its predecessor and thus submachine gun manufacture was able to keep pace with demand, even when the second front opened up with the Soviet Union in 1941. The emphasis on sub-assembly and simple manufacturing techniques meant that most small-scale engineering plants could produce the gun's various components. Both the MP38 and the MP40 are often referred to as the 'Schmeisser', but they were the products of Erma designer Heinrich Vollmer – Hugo Schmeisser was not involved at any stage.

| | |
|---|---|
| Country of origin: | Germany |
| Calibre: | 9mm Parabellum |
| Length: | 832mm (32.75in) stock extended; 630mm (24.75in) stock folded |
| Weight: | 3.97kg (8.75lb) |
| Barrel: | 248mm (9.75in), 6 grooves, rh |
| Feed/magazine capacity: | 32-round box magazine |
| Operation: | Blowback |
| Cyclic rate of fire: | 500rpm |
| Muzzle velocity: | 395mps (1300fps) |
| Effective range: | 70m (230ft) |

# Erma MP58

**In the aftermath of WWII, Erma and Walther became two of East and West Germany's most important arms manufacturers, the former producing a new range of submachine guns. The MP58 followed closely on the heels of the MP56, a wraparound-bolt weapon that had few achievements. The MP58 was produced to a Federal Government brief for an economical SMG, and Erma attempted to fulfil this by making most of the MP58 out of single-sheet steel stampings. Unfortunately, the MP58 was turned down for government adoption. This was despite the gun being essentially well made and reliable (on account of the telescopic mainspring design used during the war). Erma produced several more improved guns, but by the mid 1960s they turned away from military to commercial weapons.**

| | |
|---|---|
| Country of origin: | Germany |
| Calibre: | 9 x 19mm Parabellum |
| Length: | 405mm (16in) |
| Weight: | 3.1kg (6.8lb) |
| Barrel: | 190mm (7.5in) |
| Feed/magazine capacity: | 30-round detachable box magazine |
| Operation: | Blowback |
| Cyclic rate of fire: | 650rpm |
| Muzzle velocity: | 395mps (1295fps) |
| Effective range: | 70m (230ft) |

# Dux

The story behind the Dux is infinitely more fascinating than the features of the gun itself, it being effectively the same as the Finnish M44, but of better quality. The match with the M44 is not coincidental. In 1944, one of the M44's designers, Willie Daugs, fled to Spain from Germany and turned his M44 design drawings over to the Orviedo Arsenal. As the result of collaboration with Ludwig Vorgrimmler, the Dux 51 was born in the early 1950s and went into service with the West German Border Police as the M53, where its use continued until the end of the 1960s. Neither the M53 nor some of the redeveloped prototypes which followed found their way into military service, mainly because of legal rather than military reasons. By the 1970s, the design was being superseded and the Dux fell into disuse.

| | |
|---|---|
| **Country of origin:** | Germany/Spain |
| **Calibre:** | 9mm Parabellum |
| **Length:** | 825mm (32.48in) stock extended; 615mm (24.25mm) stock folded |
| **Weight:** | 3.49kg (7.69lb) |
| **Barrel:** | 248mm (9.75in), 6 grooves, rh |
| **Feed/magazine capacity:** | 50-round box magazine |
| **Operation:** | Blowback |
| **Cyclic rate of fire:** | 500rpm |
| **Muzzle velocity:** | 390mps (1280fps) |
| **Effective range:** | 70m (230ft) |

# Heckler & Koch MP5

**The Heckler & Koch MP5 is a masterpiece of weapons engineering. Its roller-locked delayed blowback system harks back to the German MG42 machine gun and is the same system as is used in Heckler & Koch's assault rifles. It also fires from a closed chamber, part of the reason for its considerable accuracy. The MP5 has now been in production since 1965, although the latest guns have the full range of fire-selection options: single-shot, three-round burst and full automatic. The quality of its machining is consistently excellent and much of the weapon's furniture is plastic to lighten the weapon. There are many variants of the MP5, but the two basic models are the MP5A2, which has a solid plastic butt, and the MP5A3, which has a folding metal stock.**

| | |
|---|---|
| Country of origin: | Germany |
| Calibre: | 9mm Parabellum |
| Length: | 680mm (26.77in) |
| Weight: | 2.55kg (5.62lb) |
| Barrel: | 225mm (8.85in), 6 grooves, rh |
| Feed/magazine capacity: | 15- or 30-round detachable box magazine |
| Operation: | Delayed blowback |
| Cyclic rate of fire: | 800rpm |
| Muzzle velocity: | 400mps (1312fps) |
| Effective range: | 70m (230ft) |

# Heckler & Koch MP5SD

The popularity of the Heckler & Koch MP5 series with special forces troops meant that a silenced version, known as the MP5SD, was inevitable. The 9mm Parabellum round it fires is standard and the gun's configuration is little different from any other MP5 model. Yet the integral silencer is particularly effective. The barrel of the MP5SD has 30 x 3mm holes drilled along its length and is surrounded by a two-chamber suppressor which sequentially diffuses the gases until the round leaves the muzzle at subsonic speed. Both noise and blast reduction are considerable and accuracy remains good over the reduced range. Several varieties of the MP5SD are available, each offering different configurations of furniture, fire-selection (SD 4, 5 and 6 have three-round burst facility) and sight fittings.

| | |
|---|---|
| Country of origin: | Germany |
| Calibre: | 9mm Parabellum |
| Length: | 550mm (21.65in) |
| Weight: | 2.9kg (6.39lb) |
| Barrel: | 146mm (5.75in), 6 grooves, rh |
| Feed/magazine capacity: | 15- or 30-round detachable box magazine |
| Operation: | Delayed blowback |
| Cyclic rate of fire: | 800rpm |
| Muzzle velocity: | 285mps (935fps) |
| Effective range: | 50m (164ft) |

# Heckler & Koch MP5K

The Heckler & Koch MP5K is a specially compressed version of the MP5, designed for those military or police units who need firepower which can be easily concealed until deployment. Such are its dimensions – only 325mm (12.8in) long – that it can be fitted into a briefcase, carried in a car glove compartment or held under a jacket. Its small size is achieved by the absence of a stock; instead, a front grip is provided for control. Its rate of fire is increased from that of the standard MP5 and reaches 900rpm, the high rpm giving it formidable close-range capabilities. Its compactness tends to see it only fitted with the 15-round magazine, though the 30-round magazines are perfectly useable. Four versions of the MP5K are made, including one, the MP5KA5, which has a three-round burst facility.

| | |
|---|---|
| Country of origin: | Germany |
| Calibre: | 9mm Parabellum |
| Length: | 325mm (12.8in) |
| Weight: | 2.1kg (4.63lb) |
| Barrel: | 115mm (4.53in), 6 grooves, rh |
| Feed/magazine capacity: | 15- or 30-round detachable box magazine |
| Operation: | Delayed blowback |
| Cyclic rate of fire: | 900rpm |
| Muzzle velocity: | 375mps (1230fps) |
| Effective range: | 70m (230ft) |

# Lanchester

The 9mm Lanchester was incredibly well crafted, expensive and a joy to use. It was designed for RAF and Royal Navy use (although only the Navy took receipt) after Dunkirk by George Lanchester of the Sterling Armament Company, and was almost entirely a copy of the German Bergmann MP28. Whereas the Bergmann used the furniture of the Mauser 98k, the Lanchester used that of the Short Magazine Lee-Enfield (SMLE), including the bayonet fitting. Other differences included the solid brass magazine housing, an indication of the incredible quality of materials and machining in everything from the stock to the breech-block mechanism. It fired extremely well and was loved by its owners, but its high cost and long production time meant that the crude Sten became the dominant British submachine gun.

| | |
|---|---|
| Country of origin: | Great Britain |
| Calibre: | 9mm Parabellum |
| Length: | 850mm (33.5in) |
| Weight: | 4.34kg (9.56lb) |
| Barrel: | 203mm (8in), 6 grooves, rh |
| Feed/magazine capacity: | 50-round box magazine |
| Operation: | Blowback |
| Cyclic rate of fire: | 600rpm |
| Muzzle velocity: | 380mps (1247fps) |
| Effective range: | 70m (230ft) |

# Sten Mk II

**The Sten Mk II was the definitive Sten gun, being the most numerous (more than two million produced) and having the crude metal construction that was the gun's visual signature. Its overall construction was even simpler than the Mk I's had been. The wooden trigger housing was replaced by a pressed steel box, and the butt, which was removable for cleaning the breech block and mainspring, now consisted of a simple metal tube and shoulder plate. The magazine port could also be turned to cover the aperture when the gun was not in use. The Sten Mk II was manufactured in the United Kingdom, Canada and New Zealand, and furnished not only regular Allied troops, but also many Resistance fighters who appreciated the way it could easily be broken down and concealed.**

| | |
|---|---|
| **Country of origin:** | Great Britain |
| **Calibre:** | 9mm Parabellum |
| **Length:** | 762mm (30in) |
| **Weight:** | 2.95kg (6.5lb) |
| **Barrel:** | 196mm (7.75in), 2 or 6 grooves, rh |
| **Feed/magazine capacity:** | 32-round detachable box magazine |
| **Operation:** | Blowback |
| **Cyclic rate of fire:** | 550rpm |
| **Muzzle velocity:** | 380mps (1247fps) |
| **Effective range:** | 70m (230ft) |

# Sten Mk IIS

The Sten MkIIS was an attempt to give British and Commonwealth special forces soldiers a silenced weapon for special operations, and was particularly used by RM Commandos. On the whole it was successful, as the Mk IIS was remarkably quiet, the noise of the reciprocating bolt being louder than the actual noise of firing. The Mk IIS had an integrated barrel and silencer (thus it was a separate version of the Sten altogether) and it fired bullets at a subsonic muzzle velocity of 305mps (1000fps). It was designed for single-shot use – full-automatic could be applied but created excessive wear on the baffles in the silencer. As the silencer would become extremely hot during firing, a canvas sleeve was wrapped around it as a protective foregrip. The Mk IIS later saw service during the Korean War.

| | |
|---|---|
| Country of origin: | Great Britain |
| Calibre: | 9 x 19mm Parabellum |
| Length: | 908mm (35.75in) |
| Weight: | 3.52kg (7.76lb) |
| Barrel: | 89mm (3.5in), 6 grooves, rh |
| Feed/magazine capacity: | 32-round detachable box magazine |
| Operation: | Blowback |
| Cyclic rate of fire: | 450rpm |
| Muzzle velocity: | 305mps (1000fps) |
| Effective range: | 50m (164ft) |

# Sten Mk V

By 1944, when the Sten Mk V appeared, the war had turned in the Allies' favour and a little more time and effort could be lavished on the Sten's initially crude design. The Mk V was basically the same Sten as ever in terms of operation and, crucially, feed system. The latter was significant because the Sten's magazine was the root of its unreliability; thus the Mk V carried forward that unreliability. Yet the Mk V was undoubtedly better in a cosmetic sense. It now had a wooden stock (in early models, this featured a trap for holding gun-cleaning equipment), pistol grip and, again in early models, a forward grip which was eventually dropped because it kept breaking off. It also had the muzzle and fore-sight of the Lee-Enfield No. 4 rifle and could take that rifle's bayonet.

| | |
|---|---|
| Country of origin: | Great Britain |
| Calibre: | 9mm Parabellum |
| Length: | 762mm (30in) |
| Weight: | 3.86kg (8.5lb) |
| Barrel: | 196mm (7.75in), 6 grooves, rh |
| Feed/magazine capacity: | 32-round box magazine |
| Operation: | Blowback |
| Cyclic rate of fire: | 600rpm |
| Muzzle velocity: | 380mps (1247fps) |
| Effective range: | 70m (230ft) |

# Patchett Mk 1

**The Patchett submachine gun was the brainchild of George Patchett, and it emerged as part of the British forces' attempt to develop a more satisfying replacement for the crude Sten gun. Prototypes of the weapon were produced by the Sterling Armament Co. during 1942–1943 (in 1942, Sterling had stopped making the Lanchester). These guns were promisingly tested in combat conditions during the British airborne assault at Arnhem and production of the Mk 1 started in 1944. The Patchett had a solid design and a durable blowback mechanism, and was improved to form the Patchett Mk 2, which in turn became the famous Sterling submachine gun. The main difference between the Patchett and the Sterling is that the former used a straight magazine box and could accept Sten magazines.**

| | |
|---|---|
| Country of origin: | Great Britain |
| Calibre: | 9mm Parabellum |
| Length: | 685mm (27in) |
| Weight: | 2.7kg (6lb) |
| Barrel: | 195mm (7.75in), 6 grooves, rh |
| Feed/magazine capacity: | 32-round detachable box magazine |
| Operation: | Blowback |
| Cyclic rate of fire: | 550rpm |
| Muzzle velocity: | 395mps (1295fps) |
| Effective range: | 70m (230ft) |

# Sterling L2A1

The Sterling's signature curved 34-round magazine, 50-year–plus service record and its use by more than 90 nations have made it one of the most famous firearms of the 20th century. It first entered service in the British Army in 1953 as the Sterling L2A1, with the L2A2 and L2A3 appearing in 1953 and 1956, respectively. The Sterling is a resilient and hard-working gun. Its recoil was kept under control by an advanced primer ignition system which actually fired the round a fraction of a second before the round seated itself in the chamber, the breech block then being carried backwards by the leftover force. The Sterling became a standard issue submachine gun across the world and is still made under licence in India, although Sterling ceased trading in 1988.

| | |
|---|---|
| **Country of origin:** | Great Britain |
| **Calibre:** | 9mm Parabellum |
| **Length:** | 690mm (27.16in) stock extended; 483mm (19in) stock folded |
| **Weight:** | 2.72kg (6lb) |
| **Barrel:** | 198mm (7.79in), 6 grooves, rh |
| **Feed/magazine capacity:** | 34-round detachable box magazine |
| **Operation:** | Blowback |
| **Cyclic rate of fire:** | 550rpm |
| **Muzzle velocity:** | 395mps (1295fps) |
| **Effective range:** | 70m (230ft) |

# Sterling L34A1

Development of a silenced version of the Sterling L2 began as far back as 1956, with both Patchett and the Royal Armaments Research & Development Establishment (RARDE) producting prototypes. The latter went through to acceptance and became the Sterling L34A1. The integral suppressor is very effective: the barrel has 72 radial holes drilled into it and is surrounded by a metal cylinder containing baffles into which the firing gases expand and swirl. As with all silencers, muzzle velocity is substantially reduced. Indeed, with recoil lessened, the bolt and recoil spring were lightened, thus making the weapon more manageable on full-auto than its non-silenced counterpart. Used mainly by special forces soldiers, the L34A1 provided a practical weapon offering good performance.

| | |
|---|---|
| Country of origin: | Great Britain |
| Calibre: | 9mm Parabellum |
| Length: | 864mm (34in) stock extended; 660mm (26in) stock folded |
| Weight: | 3.6kg (7.94lb) |
| Barrel: | 198mm (7.8in), 6 grooves, rh |
| Feed/magazine capacity: | 34-round detachable box magazine |
| Operation: | Blowback |
| Cyclic rate of fire: | 515rpm |
| Muzzle velocity: | 300mps (984fps) |
| Effective range: | 120m (400ft) |

# Uzi

**Beloved of Hollywood thrillers, few weapons have entered into the popular vocabulary or global service as much as the Uzi. It was designed by the talented Lieutenant Uziel Gal in the early years of Israel's existence, when Israel was desperate for a native-produced submachine gun. Gal based his design around the wrapround bolt system used in the Czech vz 23 series, in which the bolt is actually placed forward of the chamber on firing, thus saving space and allowing for a longer barrel. Gal's design was a huge success. Simply made and operated, the Uzi is easily held and packs a potent rate of fire. It initally came with a wooden stock, but now a folding metal stock is standard. Used by more than 26 countries outside of Israel, the Uzi has made a definite impact on 20th century weapons development.**

| | |
|---|---|
| **Country of origin:** | Israel |
| **Calibre:** | 9mm Parabellum |
| **Length:** | 650mm (25.6in) stock extended; 470mm (18.5in) stock folded |
| **Weight:** | 3.7kg (8.15lb) |
| **Barrel:** | 260mm (10.23in), 4 grooves, rh |
| **Feed/magazine capacity:** | 25- or 32-round box magazine |
| **Operation:** | Blowback |
| **Cyclic rate of fire:** | 600rpm |
| **Muzzle velocity:** | 400mps (1312fps) |
| **Effective Range:** | 120m (400ft) |

# Mini-Uzi

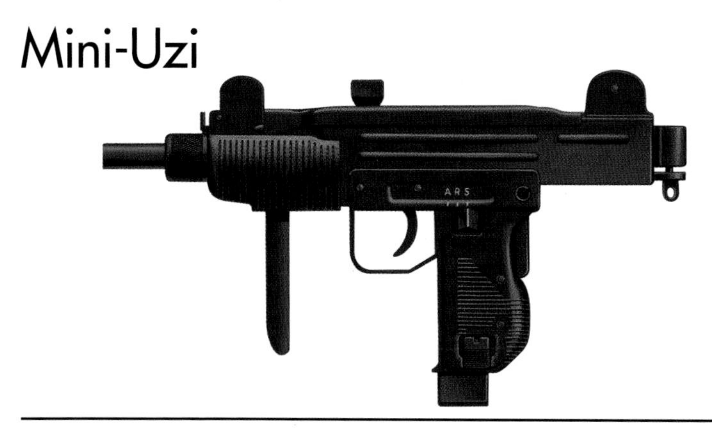


Small as the standard Uzi is in itself, the designers at Israel Military Industries have reduced it in scale not once, but twice. The mid-sized Uzi is known as the Mini-Uzi (the smallest is the Micro-Uzi) and its differences from the full-size weapon are limited almost entirely to matters of scale. Having said this, the scaling down of some components has led to a lighter bolt system, which in turn has upped the rate of fire from 600 rpm to 950rpm. This firepower can have horrifying effects at close-quarters and the Mini-Uzi has not only become prized by security organisations the world over, but also by many terrorist and criminal elements. Easily concealed, the Mini-Uzi has a special 20-round magazine (although it can accept the full-size magazines) and a single-strut stock that can be used as a foregrip when in the folded position.

| | |
|---|---|
| **Country of origin:** | Israel |
| **Calibre:** | 9mm Parabellum |
| **Length:** | 600mm (23.62in) stock extended; 360mm (14.17in) stock folded |
| **Weight:** | 2.7kg (5.95lb) |
| **Barrel:** | 197mm (7.76in), 4 grooves, rh |
| **Feed/magazine capacity:** | 20-, 25- or 32-round box magazine |
| **Operation:** | Blowback-operated |
| **Cyclic rate of fire:** | 950rpm |
| **Muzzle velocity:** | 352mps (1155fps) |
| **Effective range:** | 50m (164ft) |

# Pistole Mitragliatrice Vilar-Perosa M15

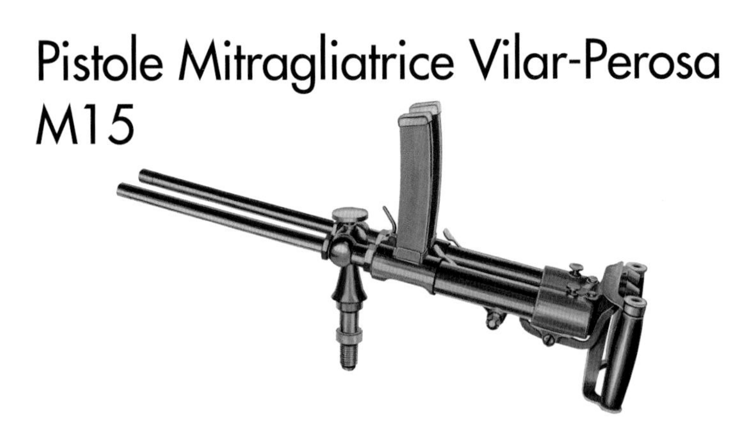

Arguably the first submachine gun despite its double barrel, the Vilar-Perosa was made as a support weapon for Alpine troops and infantry, and could be fired from the hip (suspended on straps), as well as from a bipod. Much about the weapon was advanced. Its blowback action was retarded by the breech block turning 90° following a spiral track in the receiver, the winding action imposing a slight delay on both recoil and return. However, this delay did not stop the Vilar-Perosa firing at 1200rpm, a phenomenal rate of fire which was due to a very light bolt and the top-mounted magazines, which fed the cartridges by both spring and gravity. The Vilar-Perosa served through both world wars and was a significant contribution to weapons technology.

| | |
|---|---|
| Country of origin: | Italy |
| Calibre: | 9mm Glisenti |
| Length: | 533mm (21in) |
| Weight: | 6.25kg (14.37lb) |
| Barrel: | 318mm (12.5in), 6 grooves, rh |
| Feed/magazine capacity: | 25-round box magazine |
| Operation: | Delayed blowback |
| Cyclic rate of fire: | 1200rpm |
| Muzzle velocity: | 365mps (1200fps) |
| Effective range: | 120m (400ft) |

# OVP

Although the Vilar-Perosa can justifiably lay claim to being the world's first submachine gun, its double-barrelled configuration marred its convenience of use. An attempt to create a more plausible weapon resulted in the 9mm OVP. This was one half of the Vilar-Perosa reconfigured in a standard rifle format with the addition of a wooden stock and trigger – although there was still no furniture forward of the trigger guard. Cocking also had to be redesigned and this was done on the OVP by pulling back a sleeve which wrapped around the receiver. A double-trigger unit gave the firer the option of automatic or single-shot fire. The OVP did not have a very long service life, but it was a step along the way to the very satisfying Beretta designs which would emerge following World War I.

| | |
|---|---|
| Country of origin: | Italy |
| Calibre: | 9 x 19mm Glisenti |
| Length: | 850mm (33.5in) |
| Weight: | 3.26kg (7.19lb) |
| Barrel: | 305mm (12in), 6 grooves, rh |
| Feed/magazine capacity: | 25-round detachable box magazine (top-mounted) |
| Operation: | Delayed blowback |
| Cyclic rate of fire: | 900rpm |
| Muzzle velocity: | 380mps (1247fps) |
| Effective range: | 120m (400ft) |

# Beretta Model 1938/42

**In World War II all countries had to find a balance between quality and quantity in terms of submachine gun production. In Italy, Beretta more than succeeded in this balance by turning the excellent Beretta 1938A into a form more suited to heavy wartime manufacture. This was the Model 1938/42. It closely followed the 1938A, but simplification had removed the perforated barrel jacket and shortened the barrel slightly. In addition, the metal parts of the gun were more straightforwardly engineered from sheet steel. A dust cover was also fitted to the bolt handle to stop dust and sand intrusion (a modification made from experience in the Western desert). The Model 1938/42 would only accept standard 9mm Parabellum ammunition, but it still depended on Beretta-made magazines to feed effectively.**

| | |
|---|---|
| Country of origin: | Italy |
| Calibre: | 9mm Parabellum |
| Length: | 798mm (31.4in) |
| Weight: | 2.72kg (6lb) |
| Barrel: | 198mm (7.79in), 6 grooves, rh |
| Feed/magazine capacity: | 34-round detachable box magazine |
| Operation: | Blowback |
| Cyclic rate of fire: | 550rpm |
| Muzzle velocity: | 395mps (1295fps) |
| Effective range: | 70m (230ft) |

# Beretta Model 12

A Domenico Salza design, the Model 12 emerged in the mid-1950s and once again demonstrated Beretta's superb engineering. Although Beretta now turned to an increased use of metal stampings to reduce cost, overall quality of the weapon was retained, this time applied to the tubular receiver design that was already being used in many other firearms. The Model 12's operation was an orthodox blowback, with the use of a 'wrapround' bolt to reduce the overall gun length. The receiver was larger than normal to accommodate the bolt which surrounded the barrel. The fire selector was a push-through type, and there were two safeties: one conventional, and one below the trigger guard. Accurate and hardy, the Model 12 found most use with Italian Special Forces and also in the Middle East, Africa and South America.

| | |
|---|---|
| Country of origin: | Italy |
| Calibre: | 9mm Parabellum |
| Length: | 660mm (26in) wooden stock; 645mm (25.4in) metal stock extended; 416mm (16.4in) metal stock folded |
| Weight: | 2.95kg (6.5lb) |
| Barrel: | 203mm (8in), 6 grooves, rh |
| Feed/magazine capacity: | 20-, 30- or 40-round box magazine |
| Operation: | Blowback |
| Cyclic rate of fire: | 550rpm |
| Muzzle velocity: | 380mps (1247fps) |
| Effective range: | 120m (400ft) |

# Beretta SC70

**The SC70 was a carbine variant of the AR70 designed for greater portability and use within confined areas such as buildings and vehicles. In most respects it is little different to the standard rifle, except that it had a folding metal butt stock which could be collapsed for ease of storage when required. With the butt folded the overall length of the gun was taken down to 736mm (28.9in). The SC70 also led to a weapon, the SC70 Short, which had the main dimensions reduced, in a similar way to the Soviet AKSU-74. The SC70 Short had a 320mm (12.6in) barrel, which reduced its overall accuracy but combined with the folding metal stock made the gun very convenient for use on covert or security operations when a weapon had to have limited visible presence.**

| | |
|---|---|
| Country of origin: | Italy |
| Calibre: | 5.56 x 45mm NATO |
| Length: | 960mm (37.8in) stock extended; 731mm (28.8in) stock folded |
| Weight: | 3.8kg (8.5lb) |
| Barrel: | 452mm (17.8in) |
| Feed/magazine capacity: | 30-round detachable box magazine |
| Operation: | Gas |
| Cyclic rate of fire: | 630rpm |
| Muzzle velocity: | 962mps (3180fps) |
| Effective range: | 800m (2624ft) |

# Spectre

**Apart from its appearance, the Spectre has two unique features which make it ideal for the counter-terrorist and special forces useage for which it was originally designed in the 1980s. The first is that it has a double-action trigger, thus allowing the operator to fire the gun even when it is not cocked just by a pull of the trigger alone, a useful facility in the fluid conditions of counter-terrorist operations, where situations can change rapidly requiring instant intervention. The second is its 50-round box magazine, which uses a four-column stacking system to maximise the box capacity (30-round magazines are also used). Such qualities have taken the Spectre into use by several international security forces and should ensure its future for many years to come.**

| | |
|---|---|
| Country of origin: | Italy |
| Calibre: | 9mm Parabellum |
| Length: | 580mm (22.83in) stock extended; 350mm (13.78in) stock folded |
| Weight: | 2.9kg (6.39lb) |
| Barrel: | 130mm (5.12in), 6 grooves, rh |
| Feed/magazine capacity: | 30- or 50-round detachable box magazine |
| Operation: | Blowback |
| Cyclic rate of fire: | 850rpm |
| Muzzle velocity: | 400mps (1312fps) |
| Effective range: | 50m (164ft) |

# Type 100

**Japan only designed one submachine gun during World War II and production began in 1942. The Type 100 was a conventional blowback weapon consistently let down by its underpowered 8mm Nambu pistol round, which had a tendency to jam. Notable features included a chrome-lined barrel and a muzzle break, while modifications included a folding-stock paratrooper's gun. Production ceased in 1943 before being restarted in 1944 with the Type 100/44. This came at a time when Japan was struggling to match the US forces' scale of firepower in the Pacific, and the Type 100's rpm was taken from 400 to 800. Despite simplifying the design, production never achieved the rate needed and most Japanese in the Pacific had to rely on rifles against the US BARs, M3s, M1s and Thompsons.**

| | |
|---|---|
| Country of origin: | Japan |
| Calibre: | 8mm Nambu |
| Length: | 890mm (35in) |
| Weight: | 3.83kg (8.44lb) |
| Barrel: | 228mm (9in), 6 grooves, rh |
| Feed/magazine capacity: | 30-round box magazine |
| Operation: | Blowback |
| Cyclic rate of fire: | 450rpm (1940); 800rpm (1944) |
| Muzzle velocity: | 335mps (1100fps) |
| Effective range: | 70m (230ft) |

# BXP

The BXP is an excellent weapon for security and special forces use. It was designed in the early 1980s and produced from 1988, being simply a blowback, wraparound-bolt submachine gun based around the pistol grip/magazine receiver configuration. Where the BXP excels is in overall quality. It is extremely well-balanced and can be fired one-handed when the stock is folded forward, though in this case the stock becomes a solid foregrip. It is made mostly from stainless steel and the rust-resistant coating applied extensively doubles as a dry lubricant. The distinctive muzzle of the BXP can accept a compensator, silencer or even rifle grenade and its overall accuracy is excellent owing to a barrel length of some 208mm (8.2in).

| | |
|---|---|
| Country of origin: | South Africa |
| Calibre: | 9 x 19mm Parabellum |
| Length: | 607mm (23.9in) stock extended; 387mm (15.2in) stock folded |
| Weight: | 2.5kg (5.5lb) |
| Barrel: | 208mm (8.2in), 6 grooves, rh |
| Feed/magazine capacity: | 22- or 32-round detachable box magazine |
| Operation: | Blowback |
| Cyclic rate of fire: | 1000rpm |
| Muzzle velocity: | 370mps (1214fps) |
| Effective range: | 80m (262ft) plus |

# Star SI35

The Star SI35 was a strong foray into submachine gun design let down by its complexity of manufacture. It was developed by Bonifacio Echeverria in the 1930s as part of a series intended to give Spain its own submachine gun weaponry. It was a delayed blowback gun which had the ability to switch its rates of fire between 300rpm and 700rpm, though this could not easily be done in the heat of battle. Another unusual feature was that the bolt was held back and open once the magazine was empty, a feature more common in assault rifles. The SI35 was a sound weapon in principle and in operation, and it was trialled by Britain and the US during WWII. However, the complex engineering process it used meant that both countries rejected it for home-grown options.

| | |
|---|---|
| Country of origin: | Spain |
| Calibre: | 9 x 23mm Largo |
| Length: | 900mm (35.45in) |
| Weight: | 3.74kg (8.25lb) |
| Barrel: | 269mm (10.6in), 6 grooves, rh |
| Feed/magazine capacity: | 10-, 30- or 40-round detachable box magazine |
| Operation: | Delayed blowback |
| Cyclic rate of fire: | 300 or 700rpm |
| Muzzle velocity: | 410mps (1345fps) |
| Effective range: | 50m (164ft) |

# Star Z70B

Initially the most notable feature of the Z70B submachine gun was its trigger unit, a case of 1960s design innovation over-reaching itself. In its original form, the Z62, the trigger was in two horizontal sections – pull the top section and you get automatic fire, pull the bottom section and single shots are issued. The design seemed good on paper but was unreliable. It was therefore discarded and changed to a conventional trigger unit, the gun now designated as the Z70B. The Z70B was in most other respects a normal blowback submachine gun made from metal pressings and plastic. In terms of operation, however, it was unusual in that a hammer was used to fire the rounds, controlled by the action of the bolt. The Z70B modification also brought in improved safety features.

| | |
|---|---|
| Country of origin: | Spain |
| Calibre: | 9 x 19mm Parabellum |
| Length: | 700mm (27.56in) stock extended; 480mm (18.9in) stock folded |
| Weight: | 2.87kg (6.33lb) |
| Barrel: | 200mm (7.87in), 6 grooves, rh |
| Feed/magazine capacity: | 20-, 30- or 40-round detachable box magazine |
| Operation: | Blowback |
| Cyclic rate of fire: | 550rpm |
| Muzzle velocity: | 380mps (1247fps) |
| Effective range: | 50m (164ft) plus |

# Carl Gustav M/45

Built for the Swedish Army in the immediate aftermath of World War II, the Kulspruta Pistol M/45 (as it is formally known – the Carl Gustav title relates to the factory where it was produced) is an automatic-fire, blowback weapon which is well machined and known for its reliability and durability. These qualities took it into service with the US special forces during the Vietnam War, when it was fitted with an integral silencer, and Indonesia and Egypt have also become major customers (the latter making the Carl Gustav under licence as the 'Port Said'). The stability of the original design has meant that modifications have been minimal and the current 9mm round is claimed by Sweden to be the most powerful submachine gun round available.

| | |
|---|---|
| Country of origin: | Sweden |
| Calibre: | 9mm Parabellum |
| Length: | 808mm (31.81in) stock extended; 552mm (21.73in) stock folded |
| Weight: | 3.9kg (8.6lb) |
| Barrel: | 213mm (8.38in), 6 grooves, rh |
| Feed/magazine capacity: | 36-round box magazine |
| Operation: | Blowback |
| Cyclic rate of fire: | 600rpm |
| Muzzle velocity: | 410mps (1345fps) |
| Effective range: | 120m (400ft) |

# Fürrer MP41/44

The Fürrer MP41/44's awkward appearance hints at the impracticality and inefficiency of its internal workings. It was a rushed adoption by the Swiss Army in 1940, which needed a submachine gun for its forces. Not only was this after the army rejected a competent SIG weapon, but also the MP41/44 did not emerge until three years later because of production problems. Behind the MP41/44 was the figure of Colonel Fürrer of the Federal Arms Factory, a man obsessed with applying Maxim's toggle lock to all manner of weapons from submachine guns to artillery. The result in the MP41's case was an extraordinarily complex gun which was unreliable, manufactured in insufficient numbers and uncomfortable to use. Frustrated, the Swiss Army ended up turning to Hispano-Suiza for its SMGs.

| | |
|---|---|
| **Country of origin:** | Switzerland |
| **Calibre:** | 9mm Parabellum |
| **Length:** | 775mm (30.5in) |
| **Weight:** | 5.2kg (11.5lb) |
| **Barrel:** | 247mm (9.72in), 6 grooves, rh |
| **Feed/magazine capacity:** | 40-round detachable box magazine |
| **Operation:** | Recoil, toggle-locked |
| **Cyclic rate of fire:** | 800rpm |
| **Muzzle velocity:** | 395mps (1295fps) |
| **Effective range:** | 70m (230ft) |

# SIG MP41

The SIG MP41 actually began its development life in two earlier submachine guns, the MK33 and MK37. These guns of the 1930s made little commercial impact, though they did allow SIG to experiment with different types of blowback mechanism suited to the submachine gun form. In 1941 SIG produced the MP41. This operated using a straightforward blowback system and was chambered for the 9mm Parabellum cartridge. It was solidly built, perhaps too solidly, for the receiver was made from forged steel and this combined with the extensive wooden furniture made the gun heavy to wield. Yet despite its ungainly proportions, it worked well. Thus it is more puzzling that the Swiss army preferred the problematic Furrer M41/44 as its choice of weapon, and the SIG MP41 was discontinued.

| | |
|---|---|
| Country of origin: | Switzerland |
| Calibre: | 9mm Parabellum |
| Length: | 800mm (31.5in) |
| Weight: | 4.3kg (9.6lb) |
| Barrel: | 306mm (12.05in) 6 grooves, rh |
| Feed/magazine capacity: | 40-round detachable box magazine |
| Operation: | Blowback |
| Cyclic rate of fire: | 850rpm |
| Muzzle velocity: | 400mps (1312fps) |
| Effective range: | 300m (984ft) |

# Thompson M1921

The Thompson submachine gun was designed by US Army officer Brigadier-General John Tagliaferro Thompson. In 1921, it entered the commercial market and immediately won respect. Firing the .45 ACP cartridge fed from box magazines or large-capacity drum magazines, it was a powerful weapon which fired at 800rpm with a Cutts Compensator at the muzzle to keep the gun under control. Its basic operation was delayed-blowback, the delay initially provided by the Blish Hesitation Lock, which consisted of two metal blocks which slid against each other at an oblique angle. It was actually unnecessary, and many later Thompsons did away with it. Both the US police and their criminal opponents took to the Thompson gun, and it was some years before its underworld image was overcome.

| | |
|---|---|
| **Country of origin:** | USA |
| **Calibre:** | .45in M1911 |
| **Length:** | 857mm (33.75in) |
| **Weight:** | 4.88kg (10.75lb) |
| **Barrel:** | 266mm (10.5in), 6 grooves, rh |
| **Feed/magazine capacity:** | 18-, 20- or 30-round detachable box magazine; 50- or 100-round drum magazine |
| **Operation:** | Delayed blowback |
| **Cyclic rate of fire:** | 800rpm |
| **Muzzle velocity:** | 280mps (920fps) |
| **Effective range:** | 120m (400ft) |

# Thompson M1928

The M1928 was perhaps the best known of the Thompson submachine gun series, though it differed little from the M1921. While it was the first Thompson to enter military service (with the US Marines), manufacture of the M1928 did not reach substantial levels until the beginning of WWII when France, Britain and Yugoslavia needed reliable small arms with high rates of fire. It was a complex machined weapon with laborious production processes, but gave sterling service to many Allied troops and, famously, US criminal organisations. The M1 submachine gun was the Thompson M1928 redesigned for mass production, with a simple blowback design and the removal of the M1928's barrel cooling fins and compensator. The M1 went on to widespread use with the US Army.

| | |
|---|---|
| **Country of origin:** | United States |
| **Calibre:** | .45in M1911 |
| **Length:** | 857mm (33.75in) |
| **Weight:** | 4.88kg (10.75lb) |
| **Barrel:** | 266mm (10.5in), 6 grooves, rh |
| **Feed/magazine capacity:** | 18-, 20- or 30-round detachable box magazine; 50-or 100-round drum magazine |
| **Operation:** | Delayed blowback |
| **Cyclic rate of fire:** | 700rpm |
| **Muzzle velocity:** | 280mps (920fps) |
| **Effective range:** | 120m (393ft) |

# Reising Model 55

Developed just prior to World War II by the Harrington and Richardson Arms Co., the Reising Models 50 and 55 rejected the open-bolt blowback system of most submachine guns in favour of firing from a closed bolt, the firing pin being operated within the breech block by a complex system of levers. The M50's complexity proved its undoing. The ingress of dirt through the cocking-lever track under the fore-end could quickly stop the M50's mechanism, as Marines found out to their cost in Guadalcanal, and the Reising guns were rejected by the military and passed to non-combat security personnel and police in the USA itself. The only differences between the M50 and M55 were that the latter was designed for airborne use, having a wire stock and pistol grip (instead of a wooden butt) and no compensator.

| | |
|---|---|
| Country of origin: | USA |
| Calibre: | .45in M1911 |
| Length: | 787mm (31in) stock extended; 570mm (22.5in) stock folded |
| Weight: | 2.89kg (6.37lb) |
| Barrel: | 266mm (10.5in), 6 grooves, rh |
| Feed/magazine capacity: | 12- or 25-round box magazine |
| Operation: | Delayed blowback |
| Cyclic rate of fire: | 500rpm |
| Muzzle velocity: | 280mps (920fps) |
| Effective range: | 120m (400ft) |

# United Defense M42

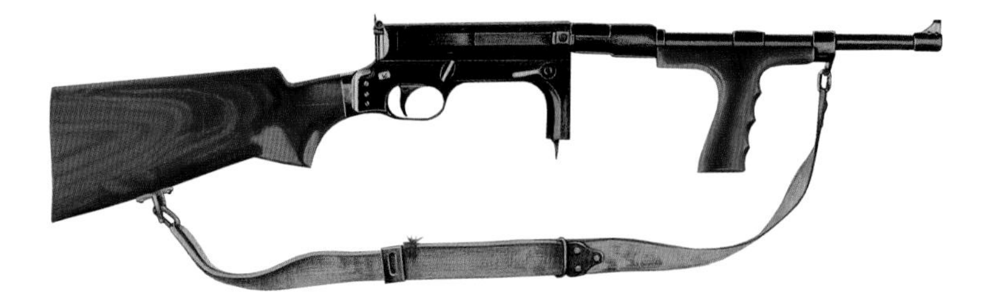

The United Defense M42 is a fine gun with an enigmatic history. It was designed by High Standard as a commercial weapon just prior to World War II and produced by Marlin to orders from the governmental United Defense Supply Corporation. The activities of this shadowy organisation are little known, but they were certainly involved in bolstering special forces and secret service activities. Thus, some 15,000 M42s went to scattered European and Far Eastern destinations, where the recipients were well served by its excellent quality of machining and finishing. It had a reliable blowback operation and a bolt handle which sealed the boltway against dirt intrusion. Despite its quality, the dominance of the Thompson guns meant the UD M42 was unable to make headway into mainstream markets.

| | |
|---|---|
| Country of origin: | USA |
| Calibre: | 9mm Parabellum |
| Length: | 820mm (32.25in) |
| Weight: | 4.11kg (9.06lb) |
| Barrel: | 279mm (11in), 6 grooves, rh |
| Feed/magazine capacity: | 20-round detachable box magazine |
| Operation: | Blowback |
| Cyclic rate of fire: | 700rpm |
| Muzzle velocity: | 400mps (1312fps) |
| Effective range: | 120m (400ft) |

# M3A1

Like so many submachine guns of World War II, the M3 was designed specifically to meet the requirements of mass production. Designed by George Hyde and produced by General Motors, the M3 came on line in 1942 and earned the title 'Grease Gun' on account of its crude appearance. It was capable of firing both 9mm and .45 ACP calibre rounds by simple changes of bolt, barrel and magazine, but the latter calibre was by far the most popular. It fired at a relatively low 450rpm via a simple blowback mechanism and, apart from one or two problems with the straight-line magazine feed and the durability of some of its cheap pressed-steel components, it proved itself a totally serviceable weapon. The M3A1's main virtue as a gun was that it was there – 650,000 were made by 1944 – and it worked.

| | |
|---|---|
| Country of origin: | USA |
| Calibre: | .45 ACP or 9mm Parabellum |
| Length: | 762mm (30in) stock extended; 577mm (22.75in) stock folded |
| Weight: | 3.7kg (8.15lb) |
| Barrel: | 203mm (8in), 4 grooves, rh |
| Feed/magazine capacity: | 30-round detachable box magazine |
| Operation: | Blowback |
| Cyclic rate of fire: | 450rpm |
| Muzzle velocity: | 275mps (900fps) |
| Effective range: | 50m (164ft) |

# Ingram M10

**The Ingram M10's notoriety relies on its brief, concealable dimensions and its astonishingly high rate of fire – more than 1000rpm. Despite this combination, the gun is surprisingly controllable, due in the main to its quality workmanship and good balance centred on the pistol grip/magazine receiver. Its designer, Gordon B. Ingram, designed the M10 from 1965–1967 in .45 ACP, although it became more commonly chambered for the 9mm Parabellum. It was originally intended to use the bulbous Sionics Company suppressor (the gun was designated the M11 in this form and used the subsonic .380 (9mm Short) calibre round), which almost reduced noise to just that of the oscillating bolt. With or without the suppressor, the M10 spread across the world amongst covert forces, special police agencies and many criminal elements.**

| | |
|---|---|
| Country of origin: | USA |
| Calibre: | .45 ACP or 9mm Parabellum |
| Length: | 548mm (21.57in) stock extended; 269mm (10.59in) stock folded |
| Weight: | 2.84kg (6.25lb) |
| Barrel: | 146mm (5.75in), 6 grooves, rh |
| Feed/magazine capacity: | 32-round box magazine |
| Operation: | Blowback |
| Cyclic rate of fire: | 1145rpm |
| Muzzle velocity: | 366mps (1200fps) |
| Effective range: | 70m (230ft) |

# 5.56mm Colt XM177E2 Commando

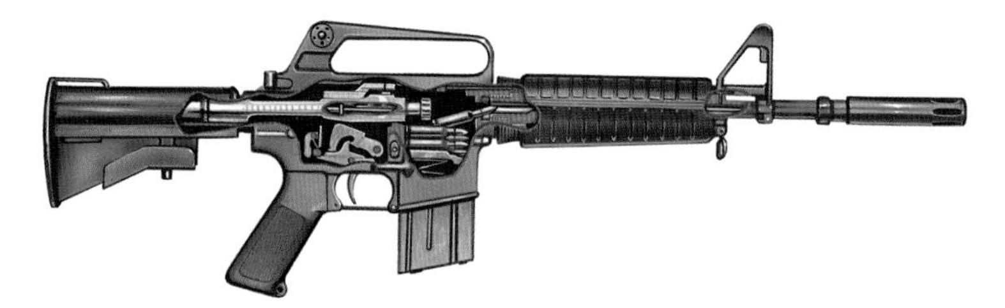

**Quite simply a shorter and lighter version of the M16 assault rifle, the XM177E2 Commando was designed specifically for those units who would value a more compact and lighter weapon in the field which still retained a high rate of fire and a powerful cartridge. Everything about the Commando apart from its size is almost identical to the M16; the parts are interchangeable and it can still be fitted with the M203 grenade launcher if required. The abbreviated dimensions have naturally made it popular with such groups as the US and worldwide special forces and security units because it is easily concealed; its commercial model is known as the Model 733. Further improvements to the rear sights enable the Commando to be used at ranges similar to those of its full-sized brother.**

| | |
|---|---|
| Country of origin: | USA |
| Calibre: | 5.56mm NATO |
| Length: | 808mm (31.81in) stock extended; 552mm (21.73in) stock folded |
| Weight: | 2.5kg (5.6lb) |
| Barrel: | 368mm (14.49in), 6 grooves, rh |
| Feed/magazine capacity: | 20- or 30-round box magazine |
| Operation: | Gas-operated self-loading |
| Cyclic rate of fire: | 700–1000rpm |
| Muzzle velocity: | 830mps (2720fps) |
| Effective range: | 400m (1312ft) |

# PPD-34/38

In the PPD-34/38, we get a foretaste of the superb PPSh-41 in terms of layout and feed. Yet the PPD weapons were actually Degtyarev's amalgamation of three existing non-Soviet weapons: the Finnish Suomi m/1931 and the German MP18 and MP28. From the m/1931 came the 71-round drum magazine, but the PPD also came with a 25-round curved box. From the German weapons came most of the PPD's actual operating system, a simple blowback. As a result of the generally high (and ultimately too expensive) standard of production and parts that went into the PPD weapons, all those made between 1934 and 1940 were generally sound weapons. This quality was nothing but enhanced by the chromium plating of the barrel to resist wear, a feature which was carried forward into the PPSh-41.

| | |
|---|---|
| Country of origin: | USSR/Russia |
| Calibre: | 7.62 x 25mm Soviet |
| Length: | 780mm (30.71in) |
| Weight: | 3.76kg (8.25lb) |
| Barrel: | 272mm (10.75in), 4 grooves, rh |
| Feed/magazine capacity: | 25-round detachable box magazine or 71-round drum |
| Operation: | Blowback |
| Cyclic rate of fire: | 800rpm |
| Muzzle velocity: | 500mps (1640fps) |
| Effective range: | 100m (328ft) plus |

# PPD-40

Following the PPD-34/38 came the last in the line of the PPD weapons – the PPD-40. In terms of performance, the gun was identical to the PPD-34/38, giving the same rpm and the same muzzle velocity, with good reliability and resistance to dirt and mishandling. The PPD-40, however, steered away from the complex machining processes of its predecessor and was altogether an easier weapon to manufacture. It also accepted the magazine through a hole set into the forestock instead of snapping into a channel. Although production had been simplified, it still fell short of the rapid stamping and pressing techniques that were demanded under wartime exigencies and thus the entry of weapons such as the PPSh-41 pushed the PPD-40 out of production.

| | |
|---|---|
| Country of origin: | USSR/Russia |
| Calibre: | 7.62 x 25mm Soviet |
| Length: | 777mm (30.6in) |
| Weight: | 3.7kg (8.16lb) |
| Barrel: | 269mm (10.6in), 4 grooves, rh |
| Feed/magazine capacity: | 25-round detachable box magazine or 71-round drum |
| Operation: | Blowback |
| Cyclic rate of fire: | 800rpm |
| Muzzle velocity: | 500mps (1640fps) |
| Effective range: | 100m (328ft) plus |

# PPSh-41

The PPSh-41 was designed to meet the urgent need for submachine guns in the Soviet Union in the wake of the German invasion in 1941. Designed by Georgiy Shpagin, it had a simple blowback action and relied on processes of metal stamping for ease of production, although it also had a chromed barrel lining. More than five million were made by manufacturers ranging from industrial plants to village workshops. Loaded with either a 71-round drum or 35-round box magazine, it could fire at 900rpm with astonishing reliability. The PPSh-41 was robust, resistant to mishandling and dirt, and powerful, and therefore both Soviet and German soldiers were eager to get their hands on the weapon and it became almost a motif of Soviet resistance to the Nazi invasion.

| | |
|---|---|
| Country of origin: | USSR/Russia |
| Calibre: | 7.62mm M1930 |
| Length: | 838mm (33in) stock extended |
| Weight: | 3.64kg (8lb) |
| Barrel: | 266mm (10.5in), 4 grooves, rh |
| Feed/magazine capacity: | 35-round box or 71-round drum magazine |
| Operation: | Blowback |
| Cyclic rate of fire: | 900rpm |
| Muzzle velocity: | 490mps (1600fps) |
| Effective range: | 120m (400ft) |

# PPS-43

The PPS-43 has been rather eclipsed by its more famous relative, the PPSh-41, yet it was much its equal even though the conditions of its production meant that only one million were made, as opposed to the five million PPSh-41s. The PPS-43 was designed by Sudarev and produced in Leningrad during the seige there in 1941–44. The seige meant that the gun had to be extremely simple in manufacture and this was achieved in the PPS-42 and then the marginally variant PPS-43. The PPS weapons were entirely made of sheet steel and their lightness and folding butt meant that production continued even after the seige, as they found popularity with armoured vehicle crews. Simple as it was, the PPS-43 had a long postwar life in the Korean War and also as a model for the Finnish M/1944.

| | |
|---|---|
| Country of origin: | USSR/Russia |
| Calibre: | 7.62 x 25mm Soviet |
| Length: | 889mm (35in) stock extended; 635mm (25in) stock folded |
| Weight: | 3.36kg (7.4lb) |
| Barrel: | 254mm (10in), 4 grooves, rh |
| Feed/magazine capacity: | 35-round detachable box magazine |
| Operation: | Blowback |
| Cyclic rate of fire: | 650rpm |
| Muzzle velocity: | 500mps (1640fps) |
| Effective range: | 100m (328ft) plus |

# Kalashnikov AKS-74U

**This short, powerful weapon is essentially a compressed version of the AKS-74 assault rifle. It has little to distinguish it from that gun except for the dramatically shortened barrel and gas tube, and the flared muzzle attachment which hides some of the flash and controls some of the recoil from the rifle-calibre round. The AKS-74U's original customers were reportedly Soviet special forces – indeed, the first use of the gun was in Afghanistan in 1982 – and also armoured vehicle personnel who needed a short weapon to store in their vehicles. This storage would have been made easier by the metal stock, which lays alongside the receiver when folded. Whoever the intended users, the AKS-74U's distribution seems to have widened to general troops.**

| | |
|---|---|
| Country of origin: | USSR/Russia |
| Calibre: | 5.45mm Soviet |
| Length: | 750mm (29.5in) stock extended; 527mm (20.7in) stock folded |
| Weight: | 3.4kg (7.5lb) |
| Barrel: | 269mm (10.6in), 4 grooves, rh |
| Feed/magazine capacity: | 30-round detachable box magazine |
| Operation: | Blowback |
| Cyclic rate of fire: | 700rpm |
| Muzzle velocity: | 488mps (1600fps) |
| Effective range: | 250m (820ft) |

# Skoda M1909

In 1888, Major George Ritter von Dormus and Archduke Karl Salvator won patents for a distinctive delayed-blowback gun, the delay provided by a pivoting block and large coil spring housed in the receiver. Following several models, the M1909 was a competitive response to the more efficient Schwarzlose machine guns produced within Austria. Its rate of fire was improved by the addition of an oiling system for the cartridge belt feed and the removal of the pendulous rate reducer which had previously kept fire to a maximum of 250rpm. The M1909 was belt-fed from a 250-round cartridge belt, as opposed to earlier gravity-fed models, but the overall inefficiencies of the series were never overcome and, after one more variation, the M1913, Skoda ceased machine gun production.

| | |
|---|---|
| Country of origin: | Austria-Hungary |
| Calibre: | 8 x 50R Austrian Mannlicher |
| Length: | 1070mm (42in) |
| Weight: | 44kg (20lb) |
| Barrel: | 525mm (20.75in) |
| Feed/magazine capacity: | 250-round fabric belt |
| Operation: | Delayed blowback, water-cooled |
| Cyclic rate of fire: | 425rpm |
| Muzzle velocity: | 618mps (2030fps) |
| Effective range: | 1000m (3300ft) plus |

# Schwarzlose M07/12

The Schwarzlose machine guns used a delayed-blowback action, in which the breech block was operated by breech pressure alone. This necessitated a high block weight and a toggle system which retarded recoil until the pressure reached levels safe enough to open the breech. The need to control pressure also led to a short barrel (the round had to leave the muzzle before breech opening), so effective range was comparatively limited at around 1000m (3300ft). Yet the range was capable enough and the gun entered service with the Austro-Hungarian Army in 1905 and several other European armies during the pre– and post–WWI periods. It emerged in four main designs, the most popular of which was the M07/12, which could also be chambered for both 7.92mm German and 6.5mm Dutch calibres.

| | |
|---|---|
| Country of origin: | Austria |
| Calibre: | 8 x 56R Austrian Mannlicher |
| Length: | 1070mm (42in) |
| Weight: | 20kg (44lb) |
| Barrel: | 525mm (20.75in), 4 grooves, rh |
| Feed/magazine capacity: | 250-round cloth belt |
| Operation: | Delayed blowback, water-cooled |
| Cyclic rate of fire: | 425rpm |
| Muzzle velocity: | 618mps (2030fps) |
| Effective range: | 1000m (3300ft) plus |

# FN MAG

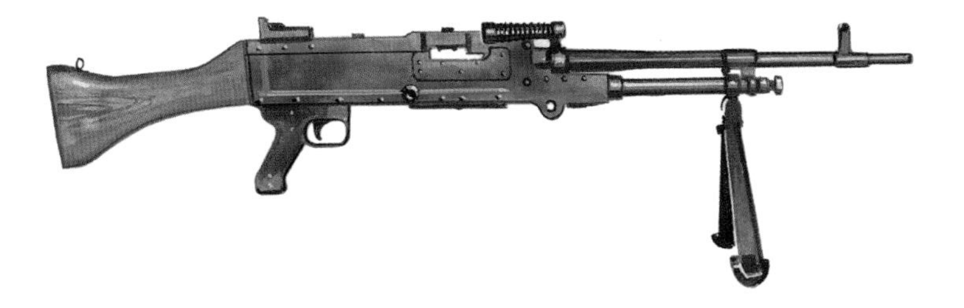

The unquestionable talents of Belgium's Fabrique Nationale Herstal are seldom more apparent than in the FN MAG (Mitrailleur à Gaz), one of the 20th century's seminal machine guns. The FN MAG perfected the concept of the general-purpose machine gun (GPMG). To adapt to different roles, a gas regulator control allows the operator to vary the gas pressure drawn from the barrel and thus alter the rate of fire. The barrel is quickly changed in combat and the whole weapon is stocky and robust, giving it a reliability that has transported it across the world. In Britain, it is made under licence as the L7A2 and more than 80 other countries have also taken it into their forces. The FN MAG can be mounted on both its own front bipod for light support roles or a heavy tripod for sustained fire deployments.

| | |
|---|---|
| Country of origin: | Belgium |
| Calibre: | 7.62 x 51mm NATO |
| Length: | 1250mm (49.2in) |
| Weight: | 10.15kg (22.25lb) |
| Barrel: | 546mm (21.5in), 4 grooves, rh |
| Feed/magazine capacity: | Metal-link belt, various lengths |
| Operation: | Gas, air-cooled |
| Cyclic rate of fire: | 600–1000rpm |
| Muzzle velocity: | 853mps (2800fps) |
| Effective range: | 3000m (9900ft) |

# FN Minimi

**The FN Minimi is a superb example of a light squad automatic weapon developed specifically for using the 5.56mm NATO round. It emerged from the FN factories in the early 1970s and its keynotes were reliability – it uses a roller-guided locking system which is extremely smooth and uncomplicated – and its ability to switch between belt feed and magazine feed (using the standard M16 magazines) without adjustment. These features and others such as a feed indicator (showing the number of rounds left in the magazine) and detachable trigger guard for gloved/NBC use have bought it a place in many armies, including the US Army, where it is known as the M249 Squad Automatic Weapon. FN also make a short barrel and telescoping stock version for special unit use.**

| | |
|---|---|
| Country of origin: | Belgium |
| Calibre: | 5.56 x 45mm NATO |
| Length: | 1040mm (40.56in) |
| Weight: | 6.83kg (15.05lb) |
| Barrel: | 466mm (18.34in), 6 grooves, rh |
| Feed/magazine capacity: | 100- or 200-round belt or 30-round magazine |
| Operation: | Gas, air-cooled |
| Cyclic rate of fire: | 750–1100rpm |
| Muzzle velocity: | 915mps (3000fps) |
| Effective range: | 2000m (6600ft) plus |

# M240

The Belgian FN MAG was, and remains, one of the most successful machine guns of the post-World War II era. Hence it was no surprise that the US armed forces adopted the weapon from the late 1970s, largely to replace the troubled M60 as a general-purpose machine gun. Relabelled the M240, it was first applied as a coaxial tank gun, but during the 1980s its use spread into other vehicular and infantry applications. As with the FN MAG, it is a 7.62 x 51mm NATO belt-fed, gas-operated weapon. Its rate of fire can be adjusted from a low (650rpm) setting to a high (950rpm) setting. There are several US variants, which include the M240C, a right-side-feed gun for use on the Bradley M2/M3 or Marine Corps Light Armored Vehicle (LAV), and the M240D, configured for use aboard helicopters.

| | |
|---|---|
| **Country of origin:** | Belgium/USA |
| **Calibre:** | 7.62 x 51mm NATO |
| **Length:** | 1260mm (49.61in) |
| **Weight:** | 13kg (28.66lb) on bipod |
| **Barrel:** | 545mm (21.46in) |
| **Feed:** | Disintegrating belt |
| **Operation:** | gas |
| **Cyclic Rate of Fire:** | 650 or 950rpm |
| **Muzzle Velocity:** | (853mps (2800 fps) |
| **Effective Range:** | 1000m (3280ft) plus |

# Lehky Kulomet ZB vz30

**The Lehky Kulomet ZB vz30 was an improved version of the superb vz26. It became the eventual model for the British Bren gun after the test centre staff at Enfield could scarcely find enough superlatives to describe its performance during trials. Like the vz26, it boasted an incredibly smooth action by virtue of the long gas cylinder (which slowed the rate of fire and impressively reduced the ferocity of recoil), and its quick-change barrel and great accuracy made it popular with troops. The main differences in the vz30 from the vz26 were modifications to the firing-pin and breech-block mechanism. Both Britain and Germany produced versions of it during WWII, the German weapon being known as the MG30(t); in either its original or licensed forms, it has been produced from Spain to China.**

| | |
|---|---|
| **Country of origin:** | Czechoslovakia |
| **Calibre:** | 7.92mm Mauser and others |
| **Length:** | 1160mm (45.75in) |
| **Weight:** | 9.6kg (21.25lb) |
| **Barrel:** | 627mm (24.7in), 4 grooves, rh |
| **Feed/magazine capacity:** | 30-round box magazine |
| **Operation:** | Gas, air-cooled |
| **Cyclic rate of fire:** | 500rpm |
| **Muzzle velocity:** | 762mps (2500fps) |
| **Effective range:** | 1000m (3300ft) plus |

# Madsen Let Maschingevaer

**The Madsen light machine gun, manufactured by Dansk Industri Syndikat from 1897, achieved incredible longevity, remaining in production until 1955 and bought by about 34 countries. It was essentially an automatic version of the manually loaded Martini rifle, with a recoil-operated breech block that moved up and down over a switch plate and, in the absence of a bolt, rounds were pushed into the chamber by a dedicated rammer and extracted by a separate mechanism. Yet its theoretical limits were more than offset by its combat performance. First used in the Russo-Japanese War in 1904, it moved through a series of calibres and formats (including a belt-feed version) to suit its many customers and is generally classed as the first true light machine gun.**

| | |
|---|---|
| **Country of origin:** | Denmark |
| **Calibre:** | Various from 6.5mm to 8mm |
| **Length:** | 1145mm (45in) |
| **Weight:** | 9kg (20lb) |
| **Barrel:** | 585mm (23in), 4 grooves, rh |
| **Feed/magazine capacity:** | 25-, 30- or 40-round box magazine |
| **Operation:** | Recoil, air-cooled |
| **Cyclic rate of fire:** | 450rpm |
| **Muzzle velocity:** | 715mps (2350fps) |
| **Effective range:** | 1000m (3300ft) plus |

# Hotchkiss Mle 1914

Hotchkiss & Co. are credited with producing the world's first viable gas-operated machine gun in 1895. This gun spawned an evolving series of weapons which ultimately equipped French, Greek, British and US forces during World War I; the Mle 1914 actually became the standard light machine gun for the French Army during that conflict. Previous Hotchkiss guns were fed by the 24- or 36-round rigid metallic strip, yet the Mle 1914 worked with what was effectively a belt feed created by linking three-round strips into a 249-round belt. The Mle 1914 proved to be a reliable weapon, although, like its predecessor, the Mle 1909, was too heavy for the mobile infantry use for which it was intended. After the war, a 7mm version was produced which opened export markets in Mexico, Spain and Brazil.

| | |
|---|---|
| Country of origin: | France |
| Calibre: | 8mm Lebel |
| Length: | 1270mm (50in) |
| Weight: | 23.6kg (52lb) |
| Barrel: | 775mm (30.5in), 4 grooves, lh |
| Feed/magazine capacity: | 24- or 30-round metallic strip or 249-round strip/belt |
| Operation: | Gas, air-cooled |
| Cyclic rate of fire: | 600rpm |
| Muzzle velocity: | 725mps (2380fps) |
| Effective range: | 2000m (6600ft) plus |

# Hotchkiss M1922/26

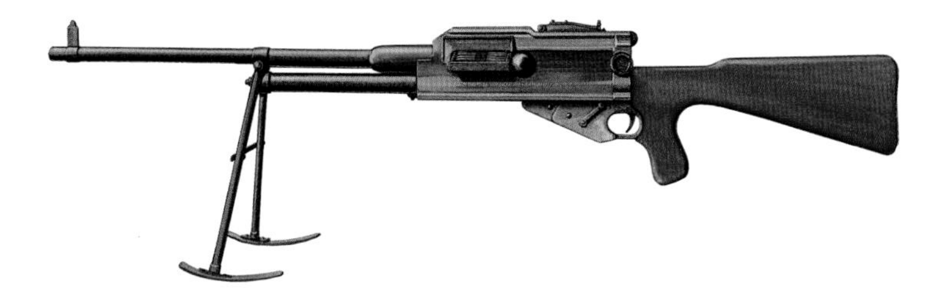

**The M1922/26 was one of several Hotchkiss models which emerged in the interwar period. On paper, several elements of the M1922/26 were promising. It had standard gas operation, but the rate of fire could be reduced through a regulator in front of the trigger. It could accept three types of feed: top-mounted magazine, the side-fed Hotchkiss metal strip or the Hotchkiss metal-link belt consisting of three-round strips joined together. The muzzle also had a climb compensator fitted. Yet the immediate aftermath of WWI was not a good time for many weapons manufacturers, and Hotchkiss struggled to make a profit. The Greek Army took some 5000 (as the M/1926) in 6.5 x 54mm Mannlicher calibre and lesser quantities went out (in 7mm calibre) to countries in South and Central America.**

| | |
|---|---|
| Country of origin: | France |
| Calibre: | 6.5mm Mannlicher and others |
| Length: | 1215mm (47.75in) |
| Weight: | 9.5kg (21lb) |
| Barrel: | 575mm (22.75in), 4 grooves, rh |
| Feed/magazine capacity: | 25- or 30-round metal strip |
| Operation: | Gas, air-cooled |
| Cyclic rate of fire: | 500rpm |
| Muzzle velocity: | 745mps (2444fps) |
| Effective range: | 1000m (3300ft) plus |

# 8mm Saint-Etienne Modèle 1907

**The Saint-Etienne Modèle 1907 (M'le'07) was a disastrous machine gun based on the equally ineffective earlier gun, the M'le'05 'Puteaux'. A textbook lesson in how to create an overcomplex, unreliable weapon, the M'le'07 reversed the gas-piston operation of its forebear to a forward direction, thus requiring a rack-and-pinion system to push the bolt backwards. This unnecessary complexity was compounded by the return spring being directly beneath the barrel, where the spring was weakened by the barrel's heat drawing the temper from the steel. Such problems made it a precarious weapon to use in dirty environments (i.e. the entire Western Front) and it was removed from service. Its one interesting feature was that adjustments could be made to the gas cylinder to vary the rate of fire.**

| | |
|---|---|
| **Country of origin:** | France |
| **Calibre:** | 8 x 50R Lebel |
| **Length:** | 1180mm (46.5in) |
| **Weight:** | 25.75kg (57lb) |
| **Barrel:** | 710mm (28in), 4 grooves, rh |
| **Feed/magazine capacity:** | 24- or 30-round metallic strip |
| **Operation:** | Gas, air-cooled |
| **Cyclic rate of fire:** | 500rpm |
| **Muzzle velocity:** | 700mps (2300fps) |
| **Effective range:** | 2000m (6600ft) |

# Fusil Mitrailleur M'15 ('Chauchat')

This cheap, roughly produced and unreliable weapon has earned itself the title of the worst machine gun in history from some authorities. The accusation seems sustainable. Known commonly as the 'Chauchat', after the commission that brought it into service, its long-recoil system was entirely unsuitable for its relatively low weight: the barrel and bolt recoiled for the full stroke before the barrel returned to its firing position (known as its 'battery') and then the bolt released itself, extracting the spent cartridge and chambering the new round. This violent action was hard to control, as was the Chauchat's tendency to jam owing to the poor quality and ill adjustment of many of its components. Used by France, Belgium and Greece during World War I, it was a precarious and hated weapon on which to rely in combat.

| | |
|---|---|
| Country of origin: | France |
| Calibre: | 8 x 50R Lebel |
| Length: | 1145mm (45in) |
| Weight: | 9kg (20lb) |
| Barrel: | 469mm (18.5in), 4 grooves, rh |
| Feed/Magazine capacity: | 20-round box magazine |
| Operation: | Recoil, air-cooled |
| Cyclic rate of fire: | 250rpm |
| Muzzle Velocity: | 700mps (2300fps) |
| Effective Range: | 1000m (3300ft) plus |

# Fusil Mitrailleur M'le 24/29 Châtellerault

The M'le 24 was France's positive escape in the late 1920s from the hideous deficiencies of the Chauchat. The awkward rimmed 8mm Lebel cartridge was replaced with a rimless 7.5mm round which fed more easily. Recoil operation gave way to gas operation, based on the US BAR (Browning Automatic Rifle). The M'le 1924 worked fairly well, although some modifications were required to the cartridge type, when the gun was redesignated the M'le 1924/29 Châtellerault, after its place of development. Produced between 1930 and 1940, it served until the 1950s and was the French Army's standard light machine gun during WWII. A static defence version for the Maginot Line featuring a 150-round side-mounted drum magazine was known as the M'le 1931, which was also used on French armour.

| | |
|---|---|
| Country of origin: | France |
| Calibre: | 7.5 x 54mm M29 |
| Length: | 1080mm (42.5in) |
| Weight: | 9.25kg (20.25lb) |
| Barrel: | 500mm (19.75in), 4 grooves, rh |
| Feed/magazine capacity: | 25-round box magazine |
| Operation: | Gas, air-cooled |
| Cyclic rate of fire: | 500rpm |
| Muzzle velocity: | 825mps (2700fps) |
| Effective range: | 1000m (3300ft) plus |

# Arme Automatique Transformable (AAT-52)

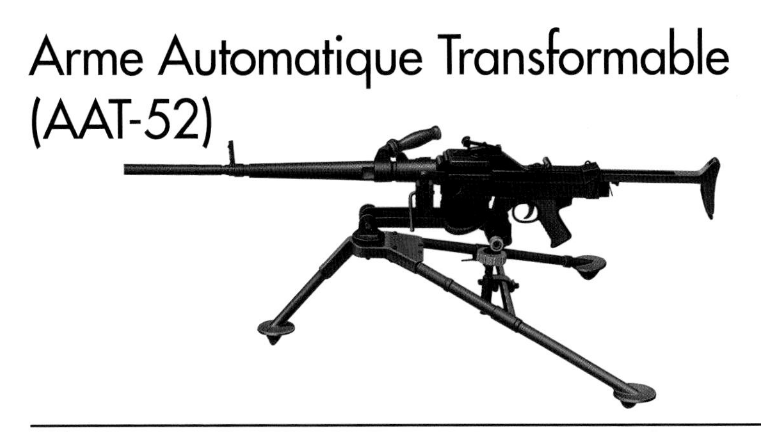

The Arme Automatique Transformable (AAT-52) was produced from 1952 and became the French Army's standard general-purpose machine gun. A rather ungainly looking weapon, it works on a delayed-blowback operation which has a two-part bolt: a lighter front section which receives the initial impetus of recoil and forces through a delay lever to push back a heavy rear section and open the breech. The AAT-52 has some good features, not least its ability to switch between light and heavy barrels for support- and sustained-fire roles, respectively. The latter role has a purpose-designed tripod for the purpose, as opposed to the standard bipod which is attached to the barrel, a real inconvenience for barrel changes. The AAT-52 was later converted to the 7.62mm NATO round and was labelled the AAT-52/mle NF-1.

| | |
|---|---|
| **Country of origin:** | France |
| **Calibre:** | 7.5 x 54mm M1929/7.62 x 51mm NATO |
| **Length:** | 990mm (38.97in) stock retracted; 1145mm (45in) stock extended |
| **Weight:** | 9.9kg (21.75lb) |
| **Barrel:** | 500mm (19.68in), 4 grooves, rh; heavy barrel 600mm (23.62in) |
| **Feed/magazine capacity:** | 50-round, disintegrating-link belt |
| **Operation:** | Delayed blowback, air-cooled |
| **Cyclic rate of fire:** | 475rpm |
| **Muzzle velocity:** | 840mps (2755fps) |
| **Effective range:** | 1000m (3300ft) |

# NF-1 GPMG

The AAT-52 was a perfectly good weapon in all respects and served French forces well. However, its 7.5 x 54mm cartridge set it at odds with NATO standardisation to the 7.62mm round, so old machine guns have been recalibrated and new made to conform with NATO specifications. In this format it is know as the NF-1. It comes in light- and heavy-barrelled versions but in both cases the switch to the 7.62 x 51mm NATO cartridge has resulted in a marked improvement in performance – the NF-1 has increased the AAT-52's effective range by 200–400m (656–1312ft). Yet some problems remain, especially the difficult barrel change arrangement in which the bipod comes off with the barrel and thus gives the operator no support while holding the hot gun.

| | |
|---|---|
| **Country of origin:** | France |
| **Calibre:** | 7.62 x 51mm NATO |
| **Length:** | 1245mm (49in) |
| **Weight:** | 11.37kg (25lb) |
| **Barrel:** | 600mm (23.62in), 4 grooves, rh |
| **Feed/magazine capacity:** | Disintegrating-link belt |
| **Operation:** | Delayed blowback |
| **Cyclic rate of fire:** | 900rpm |
| **Muzzle velocity:** | 830mps (2723fps) |
| **Effective range:** | 1200m (3937ft) plus |

# Maxim Maschinengewehr '08

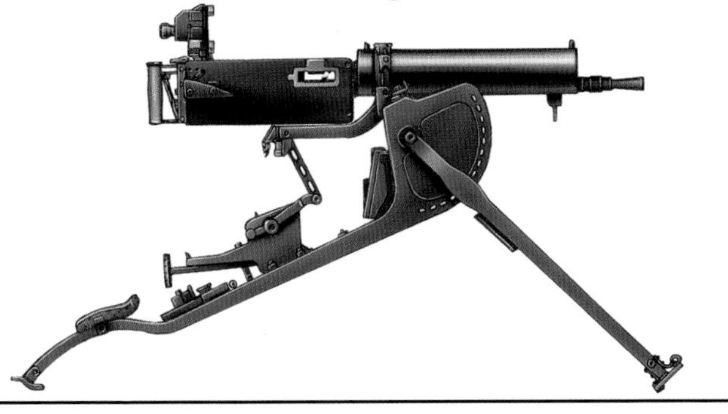

With the Vickers machine gun, the Maxim '08 dominated the landscape of the Western Front during World War I. Like the previous weapons developed by Hiram Maxim in the late 1890s, the '08 was a short-recoil gun which used a toggle system in breech locking and was water cooled. The '08 became the foremost frontline machine gun of German forces during World War I; a single gun could produce dependable fire for hour after hour at 450rpm when fitted with a muzzle booster to increase recoil force. The one drawback of the '08 was its mounted weight – 62kg (136.69lb) with its sledge mounting – and the subsequent '08/15 replaced the sledge with a bipod and shoulder stock which reduced overall weight to little more than 18kg (39.68lb).

| | |
|---|---|
| Country of origin: | Germany |
| Calibre: | 7.92 x 57mm Mauser |
| Length: | 1175mm (46.25in) |
| Weight: | 26.44kg (58.29lb) |
| Barrel: | 719mm (28.3in), 4 grooves, rh |
| Feed/magazine capacity: | 250-round belt |
| Operation: | Short recoil, water-cooled |
| Cyclic rate of fire: | 300–450rpm |
| Muzzle velocity: | 892mps (2925fps) |
| Effective range: | 2000m (6600ft) plus |

# 7.92mm Bergmann MG15Na

The Bergmann MG15 was the WWI production version of the earlier MG10, brainchild of Theodor Bergmann and Louis Schmeisser. Like many weapons in the history of German gun production, it was somewhat ahead of its time, using a modern aluminium-link belt-feed system and a short-recoil operation. Initially water cooled, it was superseded during the war by the air-cooled MG15Na (the previous water jacket was replaced by a slotted metal barrel surround) which featured a pistol grip and trigger, a recoil pad fitted to the rear of the receiver, a tripod mounting and a drum magazine to contain the belt. The MG15Na was a fine gun which served until the 1930s, but the dominance of the Maxim '08 during the war meant that it never acquired much enthusiasm from military officials.

| | |
|---|---|
| Country of origin: | Germany |
| Calibre: | 7.92 x 57mm Mauser |
| Length: | 1120mm (44in) |
| Weight: | 12.9kg (28.5lb) |
| Barrel: | 726mm (28.5in), 4 grooves, rh |
| Feed/magazine capacity: | 200-round metal-link belt |
| Operation: | Recoil, air-cooled |
| Cyclic rate of fire: | 500rpm |
| Muzzle velocity: | 890mps (2925fps) |
| Effective range: | 2000m (6600ft) plus |

# Parabellum-Maschinengewehr Modell 14

As military aircraft improved, so the demand for aircraft-mounted weaponry expanded. The Maxim '08 was too heavy for airborne use, so the engineering company Deutsche Waffen und Munitionsfabrik (DWM) was commissioned to design a lighter weapon. The Parabellum machine gun appeared in 1911 and utilised the same upward toggle break mechanism used by the Vickers machine gun and Parabellum pistol. It was very successful in its intended role, with a high rate of fire and good reliability. The MG14 was the dominant incarnation and was applied to a flexible aircraft mount in aircraft or, in a water- (as opposed to air-) cooled version, in Zeppelin airships. Towards the end of WWI, the MG14 appeared in a ground-fire role in 1918 (with a thinner barrel jacket), where it performed equally well.

| | |
|---|---|
| Country of origin: | Germany |
| Calibre: | 7.92 x 57mm Mauser |
| Length: | 1225mm (48.25in) |
| Weight: | 9.8kg (21.5lb) |
| Barrel: | 726mm (28.5in), 4 grooves, rh |
| Feed/magazine capacity: | 200-round metal link belt |
| Operation: | Recoil, air-cooled |
| Cyclic rate of fire: | 600rpm plus |
| Muzzle velocity: | 890mps (2920fps) |
| Effective range: | 2000m (6600ft) plus |

# Spandau Model 1908/15

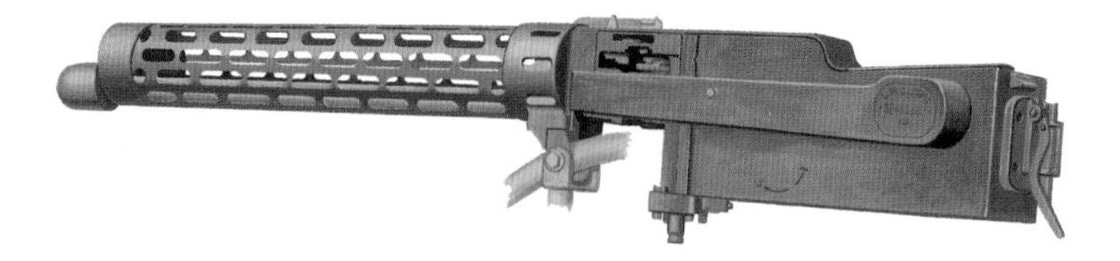

While the Maxim 08/15 was performing well on the ground, a different version was required for the expanding role of air combat during WWI. This, the Model 1908/15, was produced initially by the Spandau Arsenal and was often termed the Light Maxim. It was a conventional recoil-type Maxim, identifiable by its heavily ventilated cooling jacket which allowed effective barrel cooling when on an aircraft mount. However, the Light Maxim often had a higher rate of fire than the heavy version on account of a muzzle booster being used to increase recoil speed. Though the Spandau gun was mounted on many Fokker Eindeckers, it actually found more application as a land weapon, and after the war many were recalibrated and adapted back to water-cooling for a medium machine gun role.

| | |
|---|---|
| Country of origin: | Germany |
| Calibre: | 7.92 x 57JS |
| Length: | 1397mm (55in) |
| Weight: | 14.96kg (33lb) |
| Barrel: | 711mm (28in) 4 grooves, rh |
| Feed/magazine capacity: | Belt feed |
| Operation: | Recoil |
| Cyclic rate of fire: | 550rpm |
| Muzzle velocity: | 892mps (2925fps) |
| Effective range: | 2000m (6600ft) plus |

# MG34

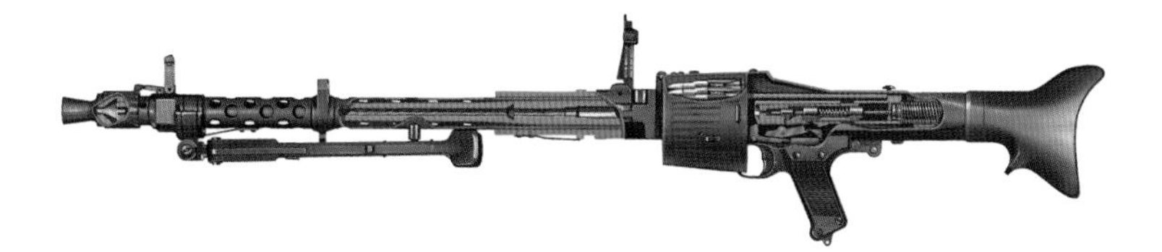

The MG34 was one of the finest machine guns of the 20th century, and effectively introduced the concept of the general purpose machine gun. It was designed to function both in assault roles using a bipod, and sustained roles on a tripod or vehicular mounting, and it excelled in both. It was also mounted as an anti-aircraft gun. The MG34 could fire at 800–900rpm, could easily be stripped simply by twisting out the barrel and was totally controllable in firing owing to the 'in-line' design which placed aligned barrel and butt horizontally. Its success took it into service as the standard German army machine gun in 1936, and it held this title until supplanted by the MG42. The reason for its replacement was purely one of cost – the MG34 was exhaustively machined and proved too expensive.

| | |
|---|---|
| **Country of origin:** | Germany |
| **Calibre:** | 7.92 x 57mm Mauser |
| **Length:** | 1219mm (48in) |
| **Weight:** | 12.1kg (26.67lb) |
| **Barrel:** | 627mm (24.75in), 4 grooves, rh |
| **Feed/magazine capacity:** | 250-round belt or 75-round saddle drum |
| **Operation:** | Recoil-operated, air-cooled |
| **Cyclic rate of fire:** | 800–900rpm |
| **Muzzle velocity:** | 762mps (2500-fps) |
| **Effective range:** | 2000m (6600ft) plus |

# MG42

The MG42 emerged in 1941 from Germany's desperate need to speed up the process of machine gun manufacture over that of the superb, but expensive and intricate, MG34. Mauser took up the challenge in 1940, and looked to the new production methods that were already pumping out weapons such as the MP40 submachine gun. The MG42's production advantage was that it used processes of metal stamping and welding instead of machining on the receiver and barrel housing, processes which dramatically increased output. A superb new locking system was introduced, using two locking rollers which cammed outwards to recesses on the receiver walls. The result was a light, accurate gun with a very high rate of fire which produced a curious rasping sound likened to tearing linoleum.

| | |
|---|---|
| Country of origin: | Germany |
| Calibre: | 7.92 x 57mm Mauser |
| Length: | 1220mm (48in) |
| Weight: | 11.5kg (25.35lb) |
| Barrel: | 535mm (21in), 4 grooves, rh |
| Feed/magazine capacity: | 50-round belt |
| Operation: | Short recoil operated, air cooled |
| Cyclic rate of fire: | 1200rpm |
| Muzzle velocity: | 800mps (2650fps) |
| Effective range: | 3000m (10,000ft) plus |

# MG42/59

Following its superb wartime performance, the MG42 continued in several new variants adapted for the new standard 7.62 x 51mm NATO round. These were the MG1, the MG3A1 and the MG42/59. The MG42/59 is actually a commercial version of the MG3 designed for export. Its dominant buyer is Italy (Austria was another client) and indeed several Italian manufacturers are now involved in its production, including Beretta and Luigi Franchi. In most senses it is no different to the MG3 in service with the German Army, but its dimensions are slightly different, which overall makes it 0.5kg lighter than the MG3. Its rate of fire is also lessened from the German model to around 800rpm. The MG42/59 is now the Italian Army's standard light machine gun and it performs equally well as any other gun in the series.

| | |
|---|---|
| Country of origin: | Germany |
| Calibre: | 7.62 x 51mm |
| Length: | 1220mm (48.03in) |
| Weight: | 12kg (26.45lb) |
| Barrel: | 531mm (20.9in), 6 grooves, rh |
| Feed/magazine capacity: | Belt |
| Operation: | Short recoil, air cooled |
| Cyclic rate of fire: | 800rpm |
| Muzzle velocity: | 820mps (2690fps) |
| Effective range: | 3000m (10,000ft) plus |

# 7.62mm Maschinengewehr 3

**The 7.62mm Maschinegewehr 3 (MG3) is instantly recognisable as a derivative of the revered MG42. During the 1950s, the German Army assessed international weaponry for its rearmament and found that the MG42 was yet to be surpassed. Thus the MG1 was produced, almost identical in every way to the MG42, with fractional changes to feed mechanism and bolt. An evolving chain of upgrades and modifications ensued (in particular, the shift to the 7.62mm NATO rounds), leading to the current MG3, the standard light machine gun used by the German forces and many other armies around the world. The MG3 retains the original MG42's high rate of fire, although this can be varied with the use of the V550 bolt or the heavier V950 bolt. The feed can equally accept German DM1 and DM13 belts and US M13 belts.**

| | |
|---|---|
| Country of origin: | Germany |
| Calibre: | 7.62 x 51mm NATO |
| Length: | 1220mm (48in) |
| Weight: | 11.5kg (25.35lb) |
| Barrel: | 531mm (20.9in), 4 grooves, rh |
| Feed/magazine capacity: | Belt feed |
| Operation: | Short recoil, air-cooled |
| Cyclic rate of fire: | 1300rpm (V550 bolt); 950rpm (V950 bolt) |
| Muzzle velocity: | 820mps (2690fps) |
| Effective range: | 3000m (9900ft) plus |

# Heckler & Koch 13E

After Heckler & Koch produced the 5.56m HK33 assault rifle in the 1970s, the company designed a new light machine gun as its partner for use as a squad support weapon. This was the HK13, and fired the same calibre round as the HK33, but from a heavier, quick-change barrel. As such it was little different to the rifle, and there soon emerged the need for some modifications. The result was the HK13E. Improvements included a longer receiver which imparted a less severe recoil, a three-round burst facility and the option of switching between magazine and belt feed with a fairly easy conversion. The furniture was also enhanced to give greater versatility to the gun. A forward grip was fitted for use when firing from the hip and also to allow the rapid change of the barrel when necessary.

| | |
|---|---|
| Country of origin: | Germany |
| Calibre: | 5.56 x 45mm NATO |
| Length: | 1030mm (40.55in) |
| Weight: | 8kg (17.64lb) |
| Barrel: | 450mm (17.72in), 6 grooves, rh |
| Feed/magazine capacity: | 20- or 30-round detachable box or belt feed |
| Operation: | Roller-locked delayed blowback, air-cooled |
| Cyclic rate of fire: | 750rpm |
| Muzzle velocity: | 925mps (3035fps) |
| Effective range: | 1000m (3300ft) plus |

# Heckler & Koch HK21

**Essentially a belt-feed, heavy-barrelled version of the ubiquitous G3 rifle, the HK21 maintained the rifle's general excellence in quality and fire capability. It fulfilled all the criteria for a general-purpose machine gun: light, with a good rate of suppressive fire (900rpm) and an effective range of around 2000m (6600ft). These qualities took it from German service to use by Portugal (where it is now manufactured under licence), Africa and Southeast Asia. As part of the Heckler & Koch range, it was easy to modify the HK21 for G3 magazine feed and it could also accept 5.56 x 45mm or 7.62 x 39mm rounds with changes of the barrel, belt-feed plate and bolt. The HK21 was followed by two subsequent designs: the HK21A1 (which had the magazine option removed and faster belt loading) and the HK21E.**

| | |
|---|---|
| Country of origin: | Germany |
| Calibre: | 7.62 x 51mm NATO |
| Length: | 1021mm (40.2in) |
| Weight: | 7.92kg (17.46lb) |
| Barrel: | 450mm (17.72in), 4 grooves, rh |
| Feed/magazine capacity: | Belt feed |
| Operation: | Delayed blowback, air-cooled |
| Cyclic rate of fire: | 900rpm |
| Muzzle velocity: | 800mps (2625fps) |
| Effective range: | 2000m (6600ft) |

# Heckler & Koch HK21E

Heckler & Koch's continual development of its HK21 range of machine guns has now arrived at the HK21E series. The appearance is little changed from the standard HK21, but the receiver has been lengthened to create a more manageable recoil, adding an extra 119mm (4.69in) to the overall length. As with the HK21, the HK21E can take both belt and magazine feed through simple adaptation of the bolt mechanism; in the case of the HK21E, firing options have been extended to include a three-round burst in addition to single shot and full automatic. Further improvements in the HK21E are drum-type rear sights (which work better with the lengthened receiver) and the fitting of front grips to allow firing from the hip in assault manoeuvres.

| | |
|---|---|
| Country of origin: | Germany |
| Calibre: | 7.62 x 51mm NATO |
| Length: | 1140mm (44.88in) |
| Weight: | 9.3kg (20.5lb) |
| Barrel: | 560mm (22.04in), 4 grooves, rh |
| Feed/magazine capacity: | Belt feed, variable belt length |
| Operation: | Gas, air-cooled |
| Cyclic rate of fire: | 800rpm |
| Muzzle velocity: | 840mps (2755fps) |
| Effective range: | 1000m (3280ft) plus |

# Maxim .45 Mk 1

**Maxim's ground-breaking machine guns entered active service with the British Army in the late 1800s, and were soon proving their military value in action in Britain's African colonies. The first gun to be adopted by the British was designated 'Gun, Maxim, 0.45in, Mk 1'. This was first calibrated for the heavy .455in calibre round that was also used in the Martini-Henry breech-loading rifle, then in the Royal Navy's 0.45in Gardner-Gatling round, but in 1889 a .303in version was established and this became the norm (though .45in guns would continue in service until 1915). The 0.303in round was awkward for Maxim in that he struggled to get enough recoil to operate the automatic action, though the change to a cordite propellant solved much of that problem.**

| | |
|---|---|
| Country of origin: | Great Britain |
| Calibre: | .303in British |
| Length: | 1180mm (46.5in) |
| Weight: | 18.2kg (40lb) |
| Barrel: | 720mm (28.25in), 4 grooves, rh |
| Feed/magazine capacity: | Belt feed |
| Operation: | recoil, water cooled |
| Cyclic rate of fire: | 600rpm |
| Muzzle velocity: | 600mps (1970fps) |
| Effective range: | 2000m (6561ft) |

# Hotchkiss Mk 1

As machine guns became in increasing demand during WWI, the British decided to manufacture the Hotchkiss Mle 1909 under licence in the .303in British calibre and relabelling it '0.303in Gun, Machine, Hotchkiss, Mk 1'. Essentially, the British gun was the same as the French original, but with several significant improvements to aid its function and deployment. Most important was the replacement of the central tripod on the French gun with a butt and a bipod. This gave the gun a much greater ease of application as a light machine gun. The strip feed, however, was still unsatisfactory, so, from mid-1917, the Mk 1* gun was issued. This was capable of switching between strip and belt feed (although the latter was actually composed of three-round strips linked together).

| | |
|---|---|
| **Country of origin:** | Great Britain |
| **Calibre:** | .303in British |
| **Length:** | 1187mm (46.73in) |
| **Weight:** | 12.25kg (27lb) |
| **Barrel:** | 596mm (23.5in), 4 grooves, rh |
| **Feed/magazine capacity:** | 30-round metal strip, or belt feed (three-strip links) |
| **Operation:** | Gas, air-cooled |
| **Cyclic rate of fire:** | 500rpm |
| **Muzzle velocity:** | 739mps (2425fps) |
| **Effective range:** | 1000m (3300ft) |

# Lewis Gun Mk 1

The Lewis gun is named after US Army Colonel Isaac Newton Lewis, even though he actually did little more than perfect a design by the physician and weapons designer Samuel McClean. Yet, having had the gun rejected by the US Army, Lewis took it to Europe where it was adopted by the Belgian Army in 1913 and manufactured under licence in Britain from 1914. Despite the intrigues and acrimony of its origins, the Lewis gun was a fine weapon with many superb innovations. Gas operated, it had a rotating-bolt system driven by the gas piston itself. It was cooled via a system of fins between the muzzle and barrel casing which drew cool air into the jacket through the air current set up by the muzzle blast. The gun proved itself to be a dependable weapon in both world wars.

| | |
|---|---|
| Country of origin: | Great Britain |
| Calibre: | .303in British Service and others |
| Length: | 965mm (38in) |
| Weight: | 11.8kg (26lb) |
| Barrel: | 666mm (26.25in), 4 grooves, lh |
| Feed/magazine capacity: | 47- or 97-round drum |
| Operation: | Gas, air-cooled |
| Cyclic rate of fire: | 550rpm |
| Muzzle velocity: | 745mps (2444fps) |
| Effective range: | 1000m (3300ft) plus |

# Vickers Mark 1 (Class C)

The .303 Vickers was a British improvement on the Maxim gun that became one of the most successful machine guns of all time. The improvements made included reduced weight through using high-quality steel and aluminium, a dramatically shortened receiver achieved by inverting break direction in the Maxim toggle lock system, and improved feed. The Vickers entered service with the British Army in 1912 and proved a reliable, powerful weapon which endured in production until 1945 and combat into the 1960s and beyond. The Mark I was the first of the Vickers series; the subsequent models were mostly mounting variations for air or armour use. Regardless of weather or conditions, the Vickers could maintain heavy fire for long periods, only let down in its early period by faulty ammunition.

| | |
|---|---|
| Country of origin: | Great Britain |
| Calibre: | .303in British |
| Length: | 1155mm (40.5in) |
| Weight: | 18kg (40lb) |
| Barrel: | 723mm (28.5in), 4 grooves, rh |
| Feed/magazine capacity: | 250-round fabric belt |
| Operation: | Recoil, water-cooled |
| Cyclic rate of fire: | 600rpm |
| Muzzle velocity: | 600mps (1970fps); later 730mps (2400fps) |
| Effective range: | 2000m (6600ft) |

# Rolls Royce MG

Rolls Royce has had a fine name in the defence industry for many years, though not in the field of weapons manufacture. The Rolls Royce machine gun which emerged in the 1940s was therefore somewhat of an exception. The development of the gun can be seen in the context of experimentations in very heavy calibre machine guns that several British arms manufacturers conducted in the 1930s. While Vickers produced a .50in calibre machine gun and BESA a 15mm weapon, Rolls Royce went on to produce a prototype heavy machine gun for use by troops and on armoured vehicles. While a solid gun using a Friberg/Kjellman type of breech-locking system and having a range of around 3000m (9842ft) the Rolls Royce MG stayed as a prototype and Rolls Royce kept its focus elsewhere.

| | |
|---|---|
| Country of origin: | Great Britain |
| Calibre: | .50in M2 |
| Length: | 1270mm (50in) |
| Weight: | 22.25kg (49lb) |
| Barrel: | 1020mm (40in) |
| Feed/magazine capacity: | Belt feed |
| Operation: | Recoil operated, air cooled |
| Cyclic rate of fire: | 1000rpm |
| Muzzle velocity: | 715mps (2350fps) |
| Effective range: | 3000m (9828ft) |

# Bren

A remarkable testament to the Bren gun's quality is that it is still used by the British Army today as the Bren Gun L4A2 (in 7.62mm NATO). The Bren originated from a series of trials in the 1930s for a replacement for the outdated Lewis. The winner was the Czech 7.92mm vz.27, adapted at the Royal Small Arms Factory at Enfield Lock to take the British rimmed .303 cartridge. The result was the 'Bren', a compound of Brno, its place of origin in Czechoslovakia, and Enfield. The Mark I was accurate and easy to maintain. Subsequent models speeded up production through simplification, but never lost the quality of the firearm. While the Bren did not have the same rpm and range as the German MG34 and MG42, its accuracy was exceptional owing to a very light recoil.

| | |
|---|---|
| Country of origin: | Great Britain |
| Calibre: | .303in British |
| Length: | 1150mm (45.25in) |
| Weight: | 10.25kg (22.5lb) |
| Barrel: | 625mm (25in), 6 grooves, rh |
| Feed/magazine capacity: | 30-round box magazine |
| Operation: | Gas-operated, air-cooled |
| Cyclic rate of fire: | 500rpm |
| Muzzle velocity: | 730mps (2400fps) |
| Effective range: | 1000m (3300ft) plus |

# BESA

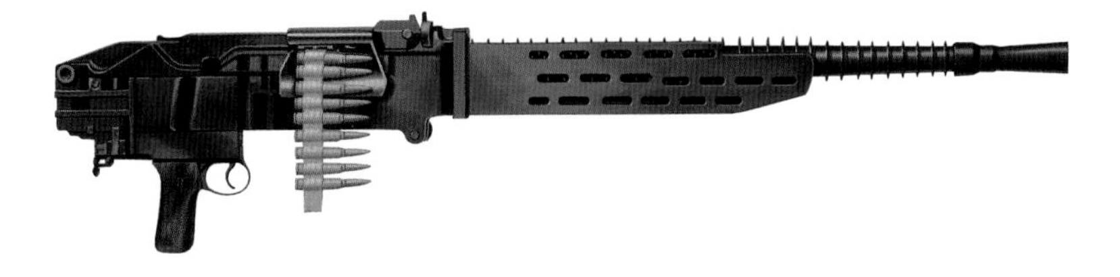

In the years leading up to and during World War II, the United Kingdom purchased two weapons from Czechoslovakia for production: the ZB26, which became the Bren gun, and the ZB53, which was the BESA when manufactured by Birmingham Small Arms (BSA) Co. The BESA was intended purely for mounting on armoured vehicles. It was gas operated, with a progressive differential recoil system in which the round was fired while the breech block was moving forwards; the discharge arrests the travel and shortens it, thus greatly reducing recoil. The BESA was thus a very accurate weapon. From the Mark 1 BESA, the gun ran through several variations; most significantly, the Mark 1 and Mark 2 had a rate selector to vary the rpm, while the Mark 3 was fixed at the high rate of fire and the Mark 3* at the low.

| | |
|---|---|
| Country of origin: | Great Britain/Czechoslovakia |
| Calibre: | 7.92mm Mauser |
| Length: | 1105mm (43.5in) |
| Weight: | 21.5kg (47lb) |
| Barrel: | 736mm (29in), 4 grooves, rh |
| Feed/magazine capacity: | 225-round belt |
| Operation: | Gas, air-cooled |
| Cyclic rate of fire: | 750–850rpm |
| Muzzle velocity: | 825mps (2700fps) |
| Effective range: | 2000m (6600ft) plus |

# BESAL Mk II

The little-known BESAL was the product of British fears during WWII that the bombing of the Royal Small Arms Factory at Enfield would severely disrupt Bren gun production. As a contingency measure, BSA were commissioned to design a low-budget version of the Bren which could be made in any reasonably equipped engineering plant if need be – the BESAL Mk I. The simplification brief was rigorously achieved: steel pressings replaced machining; the breech block and piston were of square section; the gun was cocked by pulling back the pistol grip; sights and bipod were limited to double and fixed positions, respectively. Yet, due to the soundness of the Bren design and the BESAL designer's ingenuity, the weapon performed well and would have proved a decent substitute had the need arose.

| | |
|---|---|
| **Country of origin:** | Great Britain |
| **Calibre:** | .303in |
| **Length:** | 1185mm (46.75in) |
| **Weight:** | 9.75kg (20.5lb) |
| **Barrel:** | 558mm (22in), 4 grooves, rh |
| **Feed/magazine capacity:** | 30-round box magazine |
| **Operation:** | Gas, air-cooled |
| **Cyclic rate of fire:** | 600rpm |
| **Muzzle velocity:** | 730mps (2300fps) |
| **Effective range:** | 1000m (3300ft) |

# Light Support Weapon L86A1

The introduction of the 5.56mm L85A1 as the standard British Army assault rifle meant that a new support weapon had to be developed with the same calibre. The result was the L86A1 Light Support Weapon (LSW), essentially the same weapon as the rifle, but with a heavier and longer barrel and a rear grip to aid sustained fire. The bolt system, however, is different in that it stays open when the trigger is released to allow barrel cooling. This is essential as there is no quick-change barrel, so the gun must be fired in controlled bursts with adequate time for cooling. The range of the LSW is only marginally better than the L85A1, but its accuracy is considerable; even when fitted with the standard SUSAT (Sight Unit, Small Arms Trilux) sight, it can still be used as a single-shot sniper weapon.

| | |
|---|---|
| Country of origin: | Great Britain |
| Calibre: | 5.56mm NATO |
| Length: | 900mm (35.43in) |
| Weight: | 5.4kg (11.9lb) |
| Barrel: | 646mm (25.43in), 6 grooves, rh |
| Feed/magazine capacity: | 30-round box magazine |
| Operation: | Gas, air-cooled |
| Cyclic rate of fire: | 700rpm |
| Muzzle velocity: | 970mps (3182fps) |
| Effective range: | 1000m (3300ft) |

# SIA

**Italy has produced many fine handguns and submachine guns, but its history of light and heavy machine gun manufacture is less distinguished. The SIA (Societa Anonima Italiana G. Ansaldo, Armstrong & Co.) light machine gun was produced in the 1920s and let down by several less than satisfactory features. Its delayed blowback mechanism was not the most reliable, while the magazine had an open top and bottom which only served to input dirt into the SIA's vulnerable workings and to make extraction unpredictable. However, the gun also contained some ingenuity. This included fluting the chamber gas flowing between the chamber wall and the cartridge case, which stopped the case from sticking under the firing pressure. The SIA ended up as a training weapon for the Italian Army in the 1930s.**

| | |
|---|---|
| **Country of origin:** | Italy |
| **Calibre:** | 6.5 x 52mm Carcano |
| **Length:** | Not available |
| **Weight:** | 10.66kg (25.3lb) |
| **Barrel:** | Not available |
| **Feed/magazine capacity:** | 25-round metal strip |
| **Operation:** | Delayed blowback |
| **Cyclic rate of fire:** | 700rpm |
| **Muzzle velocity:** | 645mps (2116fps) |
| **Effective range:** | Not available |

# Perino M1913

The Perino M1913 was Italy's first attempt to match the Maxim. The original gun was designed by Giuseppe Perino and patented in 1900, yet received little interest due to its excessive weight which could exceed 25kg (55.12lb). It was only with the 1913 version that the weight was reduced to 13.65kg (30lb). In other regards, however, the Perino had several salient features. Using the 6.5mm rifle cartridge, it operated through a mix of recoil and gas. Its barrel was cooled by a piston-like configuration which enclosed the barrel and pumped in cool air when the barrel was in oscillation under firing. Like many early Italian guns, its feed system tended towards over-complexity. It went from a chain feed contained in a drum to a stack magazine system with five trays holding 12 rounds each.

| | |
|---|---|
| Country of origin: | Italy |
| Calibre: | 6.5mm M95 |
| Length: | 1180mm (46.5in) |
| Weight: | 13.65kg (30lb) |
| Barrel: | 655mm (27.75in) |
| Feed/magazine capacity: | Magazine feed |
| Operation: | Combined gas/recoil, water-cooled |
| Cyclic rate of fire: | 500rpm |
| Muzzle velocity: | 740mps (2428fps) |
| Effective range: | 1500m (4950ft) |

# Fiat-Revelli Modello 14

The Modello 14 was one of Italy's first home-grown machine guns. An unusual and over-machined weapon, it served until 1945. Using the limited force of the 6.5mm M95 rifle cartridge, it had a delayed blowback operation and every round was oiled from a reservoir on top of the receiver to aid extraction (the oil attracted dust and caused stoppages). Behind the recoiling breech was an external buffer rod which impacted at 400 cycles per minute on a pad only inches from the firer's hands. This made it a fearsome weapon to use, while the tendency for cartridges to split under pressure in the chamber also made it prone to stoppages. Another distinctive feature was the feed: 10 clips of rifle ammunition set in a 10-compartment drum, a clip being fed in before the magazine revolved to the right and dropped in the next clip.

| | |
|---|---|
| Country of origin: | Italy |
| Calibre: | 6.5 x 52mm M95 |
| Length: | 1180mm (46.5in) |
| Weight: | 17kg (37.75lb) |
| Barrel: | 645mm (25.75in), 4 grooves, rh |
| Feed/magazine capacity: | 50-round strip-feed box |
| Operation: | Gas, air-cooled |
| Cyclic rate of fire: | 400rpm |
| Muzzle velocity: | 640mps (2100fps) |
| Effective range: | 1500m (4950ft) |

# Fucile Mitragliatore Breda Modello 30

The Modello 30 was one of the new breed of light machine guns that emerged from Italy in the interwar period and was an over-investment in design that came unstuck in practicality. Apart from the fact that the quick-change barrel was slightly pointless without a changing handle, the gun's defining feature was its feed system. This consisted of an integral box which was hinged on the right side of the receiver. Loading consisted of opening the box, filling it with rifle chargers, then locking it back into place. Theoretically, the design had the advantage of having more effective and dependable feed than a detachable box; however, in practice, if the magazine was damaged, the gun was useless. Add its oil-assisted extraction method and its general ungainliness, and the Modello 30 was not a great success.

| | |
|---|---|
| **Country of origin:** | Italy |
| **Calibre:** | 6.5 x 52mm M95 and others |
| **Length:** | 1230mm (48.5in) |
| **Weight:** | 10.2kg (22.5lb) |
| **Barrel:** | 520mm (20.5in), 4 grooves, rh |
| **Feed/magazine capacity:** | 20-round integral box magazine |
| **Operation:** | Blowback, air-cooled |
| **Cyclic rate of fire:** | 475rpm |
| **Muzzle velocity:** | 610mps (2000fps) |
| **Effective range:** | 1000m (3300ft) |

# Fiat-Revelli Modello 35

The Fiat-Revelli Modello 35 (M1935) was Italy's attempt to improve on the overt complexities of the Modello 14, but the result was actually to create increased unreliability and danger to the user. Out went the under-powered 6.5mm cartridge and in came an 8mm round, the chamber being fluted (grooves cut into its surface to equalise pressure inside and outside the cartridge) to cope with the power increase and dispense with the lubricating reservoir. An air-cooled system replaced the water-cooled predecessor and a standard belt feed was fitted. All these improvements were undone by major design and safety flaws. The gun fired from a closed-bolt position, which in this case led to overheating and 'cooking off' – the spontaneous firing of heated rounds. After 1945, the M1935 was scrapped.

| | |
|---|---|
| Country of origin: | Italy |
| Calibre: | 8 x 59RB |
| Length: | 1270mm (50in) |
| Weight: | 19.5kg (43lb) |
| Barrel: | 680mm (26.75in), 4 grooves, rh |
| Feed/magazine capacity: | 50-round belt |
| Operation: | Gas, air-cooled |
| Cyclic rate of fire: | 450rpm |
| Muzzle velocity: | 790mps (2600fps) |
| Effective range: | 2000m (6600ft) |

# Taisho 3

The redoubtable Hotchkiss weapons of the early 20th century were models for many a nation's machine-gun development, not least Japan's. Hotchkiss Mle 1900s made significant combat impact when used by Japanese soldiers during the Russo-Japanese War of 1904–1905 and inspired the designer Kirijo Nambu to create a similar weapon for Japanese production. Chambered for the 6.5mm Arisaka rifle cartridge, the Taisho 3 went into service in 1914 and was in most ways a simple Hotchkiss copy, although visually it was distinguished by its tripod system and the extensive use of broad cooling fins along the length of the barrel to dissipate heat. The tripod was unique in that holes sited on the tripod feet accepted poles that enabled the gun to be lifted intact by a single team.

| | |
|---|---|
| Country of origin: | Japan |
| Calibre: | 6.5 x 50mm Arisaka |
| Length: | 1155mm (45in) |
| Weight: | 28kg (62lb) |
| Barrel: | 749mm (29.5in), 4 grooves, lh |
| Feed/magazine capacity: | 30-round metal strip |
| Operation: | Gas, air-cooled |
| Cyclic rate of fire: | 400rpm |
| Muzzle velocity: | 731mps (2400fps) |
| Effective range: | 1500m (4950ft) |

# Taisho 11

World War I and the interwar years saw increased global interest in the concept of the light machine gun as an easily transportable infantry weapon; the Taisho 11 was Japan's first production of this type. It came into service in 1922 and saw action until 1945. It had several distinguishing features, such as sights offset to the right of the muzzle and receiver, and, in particular, its 30-shot, oil-lubricated hopper feed system. This ran on six clips of the same 6.5mm ammunition used in the Arisaka 38 rifle, each round being individually stripped from the clip and fed into the breech. Thus, if the machine gunner's ammunition ran out, any rifleman could provide theirs for use. The feed system was not the most reliable, however, and the oil lubrication system was problematic when dust invaded.

| | |
|---|---|
| Country of origin: | Japan |
| Calibre: | 6.5 x 50mm Arisaka |
| Length: | 1105mm (43.5in) |
| Weight: | 10.2kg (22.5lb) |
| Barrel: | 482mm (19in), 4 grooves, rh |
| Feed/magazine capacity: | 30-round hopper |
| Operation: | Gas, air-cooled |
| Cyclic rate of fire: | 500rpm |
| Muzzle velocity: | 760mps (2500fps) |
| Effective range: | 1500m (4950ft) |

# Type 89

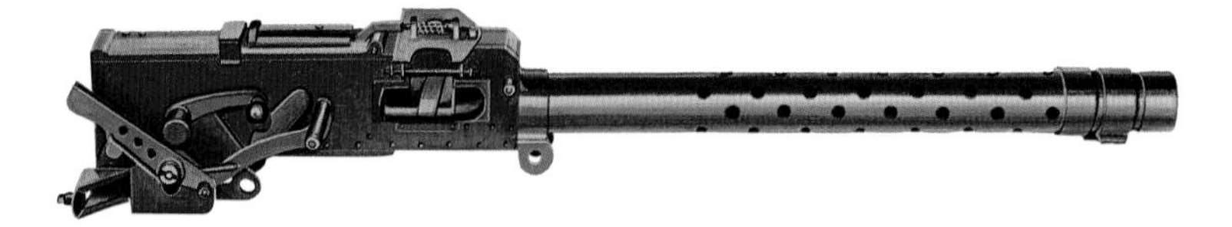

The Type 89 was a Japanese aircraft gun which found itself mounted within a variety of naval and army fighters in the Pacific theatre. Its design speaks of the tendency of Japan during WWII to copy Vickers and Browning designs and because of this, the Type 89 was one of the better weapons in Japan's arsenal at this time. It was usually mounted in pairs and was a capable gun in strafing and range-finding roles, though the latter activity was its dominant – once the pilot had ascertained target range using the Type 89 he would usually revert to his heavier cannons for the attack. Because range-finding required long, sustained bursts of fire at the target, the Type 89 was built to last, and it had a particularly heavy barrel to help disperse heat during rapid fire.

| | |
|---|---|
| Country of origin: | Japan |
| Calibre: | 7.7mm |
| Length: | 1051mm (41.4in) |
| Weight: | 16.78kg (37lb) |
| Barrel: | 685mm (27in) 4 grooves, rh |
| Feed/magazine capacity: | Fabric belt |
| Operation: | Recoil |
| Cyclic rate of fire: | 600 rpm plus |
| Muzzle velocity: | Not available |
| Effective range: | 2000m (6600ft) plus |

# Type 92

Nicknamed 'the woodpecker' by Australians who faced its stuttering fire in the Pacific, the Type 92 was issued in 1932 and signalled a shift in the Japanese Army from using the 6.5mm Arisaka round to a more potent semi-rimless 7.7mm cartridge, which increased muzzle velocity by around 32mps (100fps). The 7.7mm cartridge inexplicably came in rimmed and semi-rimmed varieties. While the rimmed cartridge would not feed through the Type 92, the other cartridges could be used without interchange modifications – although a subsequent 1941 version, known as the Type 1, could only accept the Type 99 rimless round. Apart from the ammunition, slight modifications in breech and barrel, and freedom from the need for cartridge belt oiling were all that distinguished the Type 92 from the Taisho 3.

| | |
|---|---|
| Country of origin: | Japan |
| Calibre: | 7.7mm Type 92/Type 99 |
| Length: | 1160mm (45in) |
| Weight: | 55kg (122lb) |
| Barrel: | 700mm (27.5in), 4 grooves, rh |
| Feed/magazine capacity: | 30-round metal strip |
| Operation: | Gas, air-cooled |
| Cyclic rate of fire: | 450rpm |
| Muzzle velocity: | 715mps (2350fps) |
| Effective range: | 2000m (6600ft) |

# Type 96

The Type 96 was designed as the successor to the Taisho 11, but it was not produced in sufficient numbers to supplant the latter and both saw service until 1945. Most of the Type 96's improvements over the Taisho 11 were concentrated in the feed mechanism. The cartridge oiler system was (unwisely) retained, but, instead of being located on the gun itself, it was fitted into the 30-round box magazine, itself an improvement over the Taisho 11's hopper-feed system. In addition, barrel change was made quicker. These modifications, however, were not enough to free the gun from many of the irregularities of performance and accuracy of its predecessor. Strange, therefore, that a telescopic sight was designed for the gun, as it was scarcely of value considering the capabilities of the gun.

| | |
|---|---|
| Country of origin: | Japan |
| Calibre: | 6.5 x 50mm Arisaka |
| Length: | 1055mm (41.5in) |
| Weight: | 9kg (20lb) |
| Barrel: | 555mm (21.75in), 4 grooves, rh |
| Feed/magazine capacity: | 30-round box magazine |
| Operation: | Gas, air-cooled |
| Cyclic rate of fire: | 550rpm |
| Muzzle velocity: | 730mps (2300fps) |
| Effective range: | 1000m (3300ft) |

# CETME Ameli

CETME were already renowned for their rifles when they introduced the Ameli in 1982. Its visual heritage in the German MG42 is deceptive, as the Ameli uses the roller-locked delayed blowback system it has applied in its Model L rifle, and some parts are interchangeable between the two weapons. This system makes it both reliable and fast firing, reaching up to 1200rpm, though the gun can easily be modified to reduce its rate of fire to 850rpm. The Ameli now comes in two versions – a standard weapon and a lightweight version which is over 1kg (2.2lb) lighter through its production in lighter alloys. Though currently only used by Spain and Mexico, the Ameli is a superb gun and should achieve more widespread international use.

| | |
|---|---|
| Country of origin: | Spain |
| Calibre: | 5.56 x 45mm NATO |
| Length: | 970mm (38.19in) |
| Weight: | 6.35kg (14lb) standard; 5.2kg (11.46lb) lightweight |
| Barrel: | 400mm (15.75in), 6 grooves, rh |
| Feed/magazine capacity: | 100- or 200-round boxed belt |
| Operation: | Gas operated, air cooled |
| Cyclic rate of fire: | 850 or 1200rpm |
| Muzzle velocity: | 875mps (2870fps) |
| Effective range: | 1000m (3280ft) plus |

# Solothurn MG30

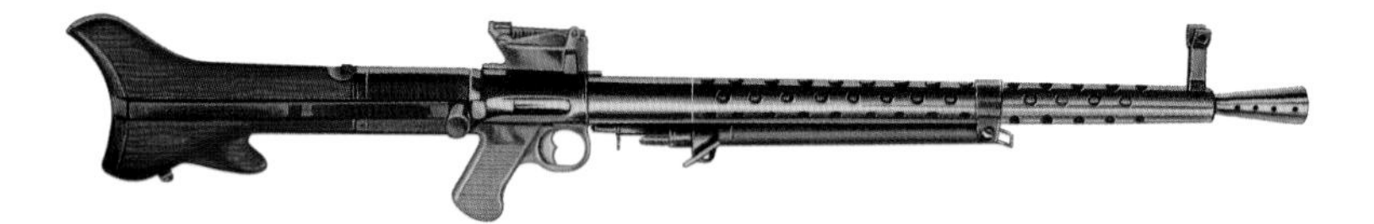

The MG30 was Solothurn's second product under the control of Rheinmetall. Though it did not achieve a great success in itself – only about 5000 were made during the 1930s – the configuration it established acted as the foundations for two of the greatest machine guns of all time, the MG34 and MG42. This heritage is visually recognisable in the MG30's straight-in-line arrangement between butt and and barrel. The MG30 had some interesting features. It was recoil-operated, worked from a side-mounted magazine, and it had a quick-change barrel facility which was operated by twisting the butt and withrawing both barrel and bolt out from the receiver. Its curious-looking rocker trigger enabled the user to fire both single shots by pulling the top section and full automatic by depressing the lower section.

| | |
|---|---|
| Country of origin: | Germany |
| Calibre: | 7.5 x 54mm Schmidt rubin |
| Length: | 1175mm (46.25in) |
| Weight: | 7.7kg (17lb) |
| Barrel: | 595mm (23.42in), 4 grooves, rh |
| Feed/magazine capacity: | 25-round detachable box magazine |
| Operation: | Recoil operated, air cooled |
| Cyclic rate of fire: | 500rpm |
| Muzzle velocity: | 800mps (2650fps) |
| Effective range: | 2000m (6561ft) plus |

# Browning M1917A1

The Browning name is legendary in the history of machine-gun development, and the M1917 started one of the finest series of firearms to date. Developed in the days prior to World War I, the .30 M1917 (so titled after its adoption date by the US Army) had much in common with Maxim guns in its water-cooled, recoil-operated configuration. The Browning was produced by several manufacturers during World War I (Remington, Colt and Westinghouse) and 68,000 were made for wartime service. The M1917A1 was a postwar (1936) revision of the weapon and differed in features of feed mechanism, sight graduation and tripod. Many other variations followed, all of which built upon the reliability and toughness of the original weapon. The M1917 was not replaced by air-cooled weapons until the late 1950s.

| | |
|---|---|
| Country of origin: | USA |
| Calibre: | .30in M1906 |
| Length: | 980mm (38.5in) |
| Weight: | 15kg (32.75lb) |
| Barrel: | 610mm (24in), 4 grooves, rh |
| Feed/magazine capacity: | 250-round fabric belt |
| Operation: | Recoil, water-cooled |
| Cyclic rate of fire: | 500rpm |
| Muzzle velocity: | 850mps (2800fps) |
| Effective range: | 2000m (6600ft) plus |

# Browning M1919A4

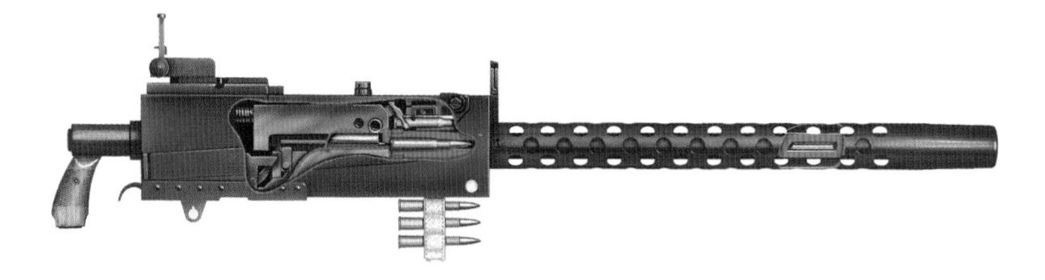

By the end of WWI a new Browning was on the drawing board as an air-cooled replacement for the water-cooled M1917. This new gun was known as the M1919, just too late to be put into active service. It would, however, take its place as one of the definitive weapons of the 20th century. The first model (M1919A1) was intended for use on armoured vehicles, but later models were designed for the cavalry (M1919A2), a general-purpose weapon (M1919A3) and, finally, an infantry version, the M1919A4. This latter gun became the prevalent form, and serves US and worldwide forces to this day. The basic M1919 configuration was a short-recoil, fabric or metal link belt fed weapon which was strong, reliable, easily controlled and capable of sustained suppressive fire beyond ranges of 2000m (6561ft).

| | |
|---|---|
| Country of origin: | United States |
| Calibre: | .30in Browning |
| Length: | 1041mm (41in) |
| Weight: | 14.05kg (31lb) |
| Barrel: | 610mm (24in), 4 grooves, rh |
| Feed/magazine capacity: | 250-round fabric or metal-link belt |
| Operation: | Recoil operated, air cooled |
| Cyclic rate of fire: | 50rpm |
| Muzzle velocity: | 853mps (2800fps) |
| Effective range: | 2000m (6561ft) plus |

# Browning M2HB

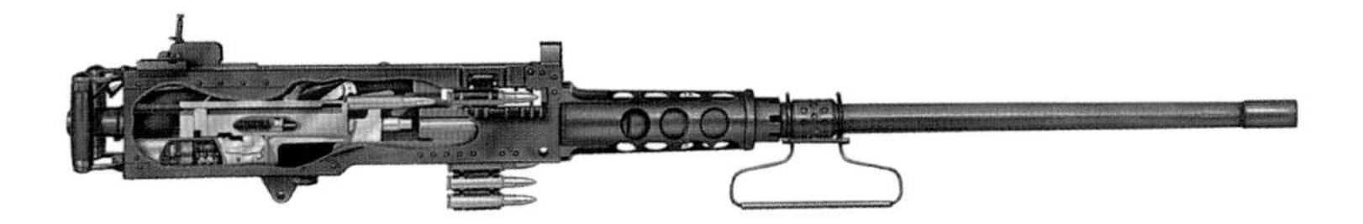

Famous the world over for the sheer power of its .50 calibre (12.7mm) round, the Browning M2HB ('Heavy Barrel') emerged from the US Army's request in 1918 for a more potent weapon with which to attack enemy vehicles and aircraft. Browning's initial answer was the water-cooled M1921 (a M1917 modified to accept Winchester's new .50 cartridge), which in the 1930s became the M2 air-cooled weapon. The cartridge's power meant that the barrel could overheat after only 75 rounds of constant firing, so a heavy barrel was added to dissipate heat and this became the dominant model, the M2HB. Apart from numerous changes to mounts and other components, the M2HB used today is essentially the same as the original. More than three million have been produced and used by the world's armies.

| | |
|---|---|
| Country of origin: | USA |
| Calibre: | .50in Browning |
| Length: | 1655mm (65in) |
| Weight: | 38.5kg (84lb) |
| Barrel: | 1143mm (45in), 8 grooves, rh |
| Feed/magazine capacity: | 110-round metallic-link belt |
| Operation: | Recoil, air-cooled |
| Cyclic rate of fire: | 450–550rpm |
| Muzzle velocity: | 898mps (2950fps) |
| Effective range: | 3000m (9900ft) plus |

# M60

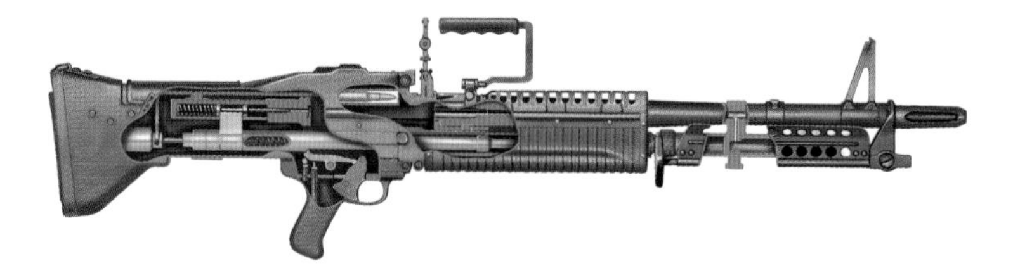

Despite being the dominant GPMG of the US Army from Vietnam to the Gulf War, the M60's history is full of problems. Developed in the late 1950s, it was an amalgam of the feed system and operation of Germany's MG42 and FG42 respectively. Despite this, and features such as the Stellite linings on the barrels which allow firing even when the barrel is white hot, the M60 was plagued by deficiencies. Barrel change was awkward because there was no handle and each barrel had its own cylinder and bipod. Thus the firer had to wrestle with the red-hot barrel using an asbestos glove. The M60E1 had its own barrel handle and kept the bipod and gas-cylinder separate from the barrel itself, but the gas-operation was prone to fouling and jamming.

| | |
|---|---|
| Country of origin: | USA |
| Calibre: | 7.62 NATO |
| Length: | 1110mm (43.75in) |
| Weight: | 10.4kg (23lb) |
| Barrel: | 560mm (22.05in), 4 grooves, rh |
| Feed/magazine capacity: | Disintegrating-link belt |
| Operation: | Gas-operated, air-cooled |
| Cyclic rate of fire: | 600rpm |
| Muzzle velocity: | 855mps (2805fps) |
| Effective range: | 3000m (9842ft) plus |

# M60E3

At 10.5kg (23.15lb), the standard M60 machine gun was a little too heavy for some light machine gun roles, so a lightweight version was produced and designated as the M60E3. The distinguishing feature of the M60E3 is the forward pistol grip which makes firing from the hip perfectly possible – although, as the gun still weighs 8.61kg (18.98lb), this act requires formidable upper body and arm strength. Other features which separate the M60E3 from its parent are a new loading mechanism for the belt and a lightweight bipod. The M60E3 has not totally overcome some of the deficiencies of its parent, but elite groups such as the US Navy SEALs and the US Marines have found it to their liking and use it in a variety of support roles.

| | |
|---|---|
| Country of origin: | USA |
| Calibre: | 7.62 x 51mm NATO |
| Length: | 1067mm (42in) |
| Weight: | 8.61kg (18.98lb) |
| Barrel: | 560mm (22.04in), 4 grooves, rh |
| Feed/magazine capacity: | Disintegrating-link belt feed |
| Operation: | Gas, air-cooled |
| Cyclic rate of fire: | 550rpm |
| Muzzle velocity: | 860mps (2821fps) |
| Effective range: | 1100m (3608ft) plus |

# Stoner M63 Machine Gun

The Stoner M63 machine gun is actually a weapons system, rather than an individual gun. It was developed by Eugene Stoner in the early 1960s based on the idea of having a standard receiver onto which different configurations of barrels, stocks and feed systems could be fitted. The result was six weapon options out of a single mechanism: carbine, light machine gun, automatic rifle, medium machine gun, mounted machine gun for vehicular use, and commando configuration (introduced later in 1969). The Stoner was a revolutionary proposition, and elite units such as the US Navy SEALs used it in Vietnam, generally with a 150-round belt box feed system. The Stoner has suffered from reliability problems, however, mainly due to its susceptibility to dirt and a weak ejector.

| | |
|---|---|
| Country of origin: | USA |
| Calibre: | 5.56 x 45mm |
| Length: | Variable; short barrel, no butt stock 660mm (25.98in) |
| Weight: | Variable; around 4.9kg (10.8lb) |
| Barrel: | Long 551mm (21.69in); short 397mm (15.6in) |
| Feed/magazine capacity: | 20-round magazine to 150-round belt |
| Operation: | Gas, air-cooled |
| Cyclic rate of fire: | 700–1000rpm |
| Muzzle velocity: | 1000mps (3300fps) |
| Effective range: | 1000m (3300ft) |

# Minigun

The M134 Minigun sits at the far edge of the category 'small arms', as it was only intended for mounted use in helicopters. Its rotating six-barrelled configuration harks back to Gatling's famous machine gun, but in this case power is supplied by an electric motor. The result is a rate of fire which can reach up to 6000rpm. This awesome firepower was brought into action specifically for the Vietnam War. Here the Minigun was clamped either into helicopter door positions or in special gun pods, and they were used for spraying the jungle floor with 7.62mm slugs. Each barrel has its own bolt unit and the feed is from a 4000-round belt which is usually stored in a drum. For sheer fire-to-size ratio, the Mingun takes some beating. It has made a number of appearances in Hollywood films.

| | |
|---|---|
| Country of origin: | USA |
| Calibre: | 7.62 x 51mm NATO |
| Length: | 800mm (31.5in) |
| Weight: | 15.9kg (35lb) |
| Barrels: | 559mm (22in), 4 grooves, rh |
| Feed/magazine capacity: | 4000-round link belt feed |
| Operation: | Electrically powered revolver |
| Cyclic rate of fire: | Up to 6000rpm |
| Muzzle velocity: | 869mps (2850fps) |
| Effective range: | 3000m (9842ft) plus |

# XM-214

The XM-214 never reached beyond the prototype stage. It followed the six-barrelled, externally powered configuration of weapons like the Minigun but used the smaller 5.56mm round which is now NATO standard. If anything, the Six-Pak, as it was affectionately know, was even more versatile in its fire suppression role than its 7.62mm predecessors. It was capable of complete control over rate of fire, as the user could select from 1000, 2000, 3000, 4000 or 6000rpm or alter the speed of barrel rotation to achieve anything between 400 and 10,000rpm. Such high rates of fire may well be the reason that the XM-214 attracted no orders, as the logistical demands of keeping such a fast-firing weapon fed with rounds were problematic. Actual feed was from two 500-round cassettes on either side of the gun.

| | |
|---|---|
| Country of origin: | USA |
| Calibre: | 5.56mm NATO |
| Length: | 685mm (27in) |
| Weight: | 38.6kg (85lb), including 1000 rounds |
| Barrel: | 455mm (18in), 4 grooves, rh |
| Feed/magazine capacity: | Belt fed from two 500-round cassettes |
| Operation: | Electrically powered revolver |
| Cyclic rate of fire: | 2000–10,000rpm |
| Muzzle velocity: | 990mps (3250fps) |
| Effective range: | 2000m (6561ft) plus |

# M249

The M249 Squad Automatic Weapon (SAW) is largely the Belgian Minimi machine gun but renamed and slightly modified for the US. Adopted by the US Army in 1982, manufacture did not actually begin until the early 1990s. Most modifications involved minor internal changes to suit American manufacturing processes, although the US gun is also fitted with a perforated steel heat shield above the barrel. This limits optical distortion from heat waves rising off the barrel. The M249 was acquired to provide base-of-fire capability for four-man infantry squads. In this role, it has more than proved itself in Afghanistan and Iraq, reliably delivering 5.56mm (0.22in) fire at a cyclical rate of 750rpm. Like the Minimi, it can feed from standard M16 magazines as well as the more typical boxed 200-round belt.

| | |
|---|---|
| Country of origin: | USA/Belgium |
| Calibre: | 5.56 x 45mm NATO |
| Length: | 1040mm (40.94in) |
| Weight: | 6.85kg (15.10lb) |
| Barrel: | 523mm (20.59in) |
| Feed: | 30-round box magazine or 200-round belt |
| Operation: | Gas |
| Cycle rate of fire: | 750rpm |
| Muzzle velocity: | 915mps (3000fps) |
| Effective range: | 1000m (3280ft) |

# Maxim MG1910

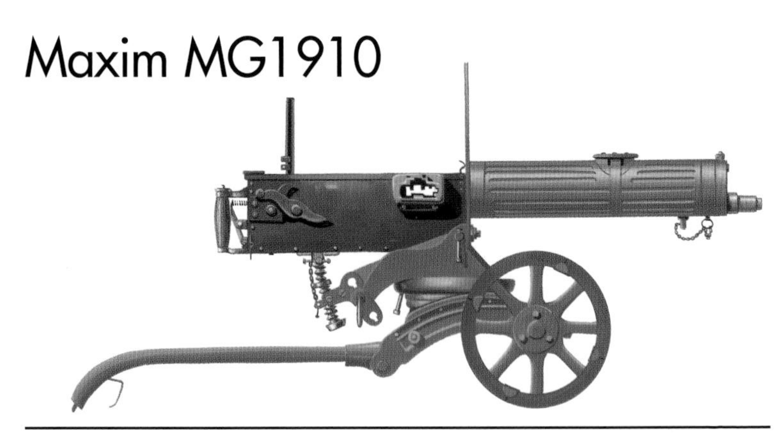

Maxim guns spread across the world either through supply or licensed manufacture. The MG1910 was the second of Russia's Maxim productions (the first being the MG1905) and was effectively a Maxim MG'08, with some modifications. Most of the improvements were concentrated into the feed mechanism which was made out of sheet steel and featured a large filling aperture at the top through which water, snow and other coolants could be quickly applied. The gun pictured here is on the standard Sokolov wheeled mount, a mount which also allowed horizontal firing action via a turntable. By mid-World War II, the Maxim was being replaced by the Goryunov SG43, but the MG1910's service history lasted well into the Cold War in many developing countries.

| | |
|---|---|
| Country of origin: | USSR/Russia |
| Calibre: | 7.62 x 55R Soviet |
| Length: | 1107mm (43.6in) |
| Weight: | 9.12kg (20.1lb) |
| Barrel: | 605mm (23.8in), 4 grooves, rh |
| Feed/magazine capacity: | 250-round fabric belt |
| Operation: | Short recoil, water-cooled |
| Cyclic rate of fire: | 550rpm |
| Muzzle velocity: | 863mps (2830fps) |
| Effective range: | 1100m (3608ft) |

# Degtyarev DP

The Degtyarev DP was a characteristically tough Russian weapon and, from 1928, its subsequent variants would stay in production until the 1950s. Its gas operation and locking system (which had locking flaps applied by the forward movement of the firing pin) were very simple and the gun offered reliability even in the filthiest of conditions. The DP was, however, slightly let down by its most prominent visual feature – its 42-round drum magazine. The flat drum was necessary to feed the awkward rimmed 7.62mm Soviet round successfully, and it could be easily damaged. Also, the gas-piston spring, being positioned underneath the barrel, was weakened over time by the heat drawing the temper out of the spring steel. Despite these problems, the DP was a strong addition to the Soviet arsenal.

| | |
|---|---|
| Country of origin: | USSR/Russia |
| Calibre: | 7.62 x 54R Soviet |
| Length: | 1290mm (50.8in) |
| Weight: | 9.12kg (20.1lb) |
| Barrel: | 605mm (23.8in), 4 grooves, rh |
| Feed/magazine capacity: | 47-round drum magazine |
| Operation: | Gas, air-cooled |
| Cyclic rate of fire: | 500–600rpm |
| Muzzle velocity: | 840mps (2760fps) |
| Effective range: | 2000m (6600ft) |

# DShK

Varieties of the DShK are still in use today around the world and nearly 50 years in production testifies to the basic soundness of its design. It was developed as a heavy machine in the early 1930s and was in demand on the Eastern Front during World War II as the DShK-38 (the number refers to year of modification). This gun used a complex rotary-feed system which was dropped after the war in favour of standard flat-feed. Although the gun itself was a good piece of engineering, the intentional weight reductions over previous Soviet machine guns were somewhat lost through its excessively heavy infantry carriage, although this could also become an anti-aircraft tripod. The DShK found its true home on armoured vehicles, especially tank turrets, and it reaches towards the Browning M2HB in performance.

| | |
|---|---|
| Country of origin: | USSR/Russia |
| Calibre: | 12.7mm Soviet |
| Length: | 1586mm (62.5in) |
| Weight: | 35.5kg (78.5lb) |
| Barrel: | 1066mm (42in), 4 grooves, rh |
| Feed/magazine capacity: | 50-round belt |
| Operation: | Gas, air-cooled |
| Cyclic rate of fire: | 550rpm |
| Muzzle velocity: | 850mps (2788fps) |
| Effective range: | 2000m (6600ft) plus |

# Goryunov SGM

The Goryunov SGM was a variant of an earlier Soviet medium machine gun, the Goryunov SG43. This had been produced in the early 1940s to replace the Maxim MG1910. The SG43 used features such as a tilting breech-block locking system and, despite an awkward feed system because of the rimmed rounds, the gun's performance was dependable. Its reliability stemmed from factors such as its very solid construction, a quick-change barrel and a chromium-plated bore. The SGM was one of a string of variants produced during WWII. Little was different to the SG43: the SGM featured a fluted barrel and a cocking handle fitted beneath the receiver. In turn, the SGM was adapted for tank use (the SGMT) and also made in a version with protective covers for its ejection and feed apertures (the SGMB).

| | |
|---|---|
| Country of origin: | USSR/Russia |
| Calibre: | 7.62 x 54R Soviet |
| Length: | 1120mm (44.1in) |
| Weight: | 13.6kg (29.98lb) |
| Barrel: | 719mm (28.3in), 4 grooves, rh |
| Feed/magazine capacity: | 250-round belt |
| Operation: | Gas, air-cooled |
| Cyclic rate of fire: | 650rpm |
| Muzzle velocity: | 850mps (2788fps) |
| Effective range: | 1000m (3300ft) |

# RPD

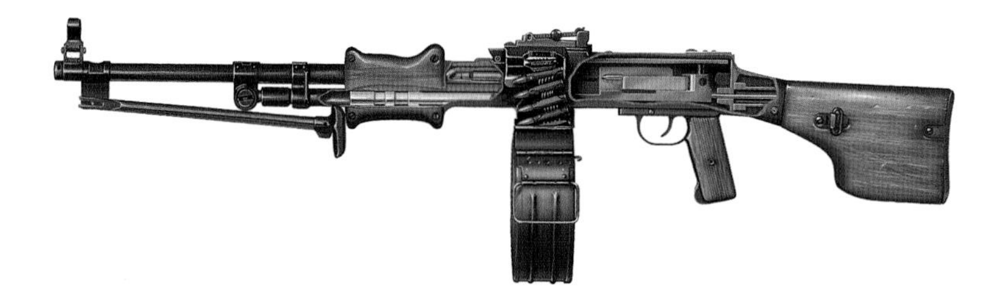

The RPD may have some hints of the Kalashnikov design in its appearance, but it is actually a Degtyarev weapon. Its abbreviation stands for Ruchnoy Pulemyot Dagtyareva and it became the standard light machine gun in the Soviet forces and those of the satellite communist states from the 1950s to the mid-1970s. The RPD had an average capability. It fired M1943 cartridges, the first Soviet machine gun to fire the intermediate round, from a drum-contained belt-feed mechanism which tended to be unreliable in dirty conditions. The RPD was strictly a light machine gun; it had no removable barrel, so fire had to be controlled and generally kept below 100rpm. Steady improvements over its 20-year service made the RPD a good squad support weapon and one which still features in the developing world.

| | |
|---|---|
| Country of origin: | USSR/Russia |
| Calibre: | 7.62 x 39mm M1943 |
| Length: | 1041mm (41in) |
| Weight: | 7kg (15.43lb) |
| Barrel: | 520mm (20.5in), 4 grooves, rh |
| Feed/magazine capacity: | 100-round belt |
| Operation: | Gas, air-cooled |
| Cyclic rate of fire: | 700rpm |
| Muzzle velocity: | 735mps (2410fps) |
| Effective range: | 900m (2952ft) |

# RPK

The RPK is quite simply a standard AKM assault rifle fitted with a longer, heavier barrel and a bipod. It was introduced as the replacement for the RPD machine gun and its identical operating method to the AKM meant that any soldier familiar with the AKM could fire the RPK without any additional training (at least in the mechanics of its firing). In some senses, however, directly scaling up the assault rifle has caused some problems. The fixed barrel means that the rpm needs to be kept to below 75 to avoid overheating – although the chromium-plating on bore and barrel make these components hard wearing. The RPK can accept any AKM magazines, but also takes a 40-round box or a 75-round drum. The more recent RPK-74 is the light support version of the AK-74 rifle, both in 5.54mm calibre.

| | |
|---|---|
| Country of origin: | USSR/Russia |
| Calibre: | 7.62 x 39mm M1943 |
| Length: | 1041mm (41in) |
| Weight: | 4.76kg (10.5lb) |
| Barrel: | 589mm (23.2in), 4 grooves, rh |
| Feed/magazine capacity: | 30- or 40-round box or 75-round drum |
| Operation: | Gas, air-cooled |
| Cyclic rate of fire: | 600rpm |
| Muzzle velocity: | 732mps (2400fps) |
| Effective range: | 800m (2600ft) |

# 7.62mm PKM

The PK established the general-purpose machine gun within the ranks of the Soviet Army and its variations equip Russian and many other armies around the world to this day. The gun is simplicity itself, being based on the Kalashnikov rotary-bolt system and having very few internal parts. Those parts work well, however, whether used in light-support or sustained-fire roles. The PK's one oddity is its use of the old M91 (1891) rimmed cartridge, which can generate some feed problems, but does outreach the M1943 round. The PK series is extensive, but each version is mainly distinguished by mountings. The PKM is rather more different, in that the weight of the barrel is reduced and there is a greater use of stampings. It and later models can be visually separated from the others by their unfluted barrel.

| | |
|---|---|
| Country of origin: | USSR/Russia |
| Calibre: | 7.62 x 39mm M1943 |
| Length: | 1160mm (45.67in) |
| Weight: | 9kg (19.84lb) |
| Barrel: | 658mm (25.9in), 4 grooves, rh |
| Feed/magazine capacity: | 100-, 200- or 250-round belt |
| Operation: | Gas, air-cooled |
| Cyclic rate of fire: | 710rpm |
| Muzzle velocity: | 800mps (2600fps) |
| Effective range: | 2000m (6600ft) plus |

# Pecheneg

The Pecheneg is a development of the PKM machine gun, chambered for 7.62x54mm. It does not have a quick-change barrel and is aimed at the squad support weapon niche rather than the general-purpose machine gun role. In service with Spetsnaz troops, the Pecheneg's non-removable folding bipod is situated much closer to the muzzle, giving the gun greater stability and accuracy of fire, while limiting the arc of fire. The carrying handle at the rear of the barrel casing is characteristically elongated. This is to protect the line of sight from possible mirages created by convection rising from the hot barrel. Earlier models used the standard flash suppressor of the PKM, which resulted in a significant muzzle blast, but current models have a more advanced suppressor that rectified this.

| | |
|---|---|
| Country of origin: | Russia |
| Calibre: | 7.62mm (.3in) |
| Length: | 1155mm (45.47in) |
| Weight: | 8.7kg (19.18lb) |
| Barrel: | 640mm (25.19in) |
| Feed/Magazine: | 100- and 200-round belt fed magazine |
| Operation: | Gas |
| Cyclic rate of fire: | 650rpm |
| Muzzle velocity: | 825m/sec (2706ft/sec) |
| Effective range: | 1500m (4921ft) |

# Steyr SSG69

When the Steyr SSG69 emerged as the Austrian Army's standard sniper rifle in 1969, it soon gathered a reputation as a well-crafted, robust and especially accurate bolt-action weapon. Equipped with the x6 Kahles ZF69 sight, it could not only achieve a guaranteed first-round kill at 800m (2624ft), but it could also then place the next 10 rounds in a grouping of less than 400mm (15.75in) at the same range (the SSG69 subsequently went on to be an internationally competitive target rifle). Being Austrian in origin, the SSG69 had to be strong enough for use by mountain troops. The bolt action is of a rear-locking Mannlicher type using six locking lugs and it uses the pre–World War I Mannlicher rotary magazine (a standard 10-round box is also applicable) – both unusual choices for a modern gun, but very strong in design. The SSG69 is fully adjustable, including stock length.

| | |
|---|---|
| Country of origin: | Austria |
| Calibre: | 7.62 x 51mm NATO |
| Length: | 1140mm (44.8in) |
| Weight: | 3.9kg (8.6lb) |
| Barrel: | 650mm (25.6in), 4 grooves, rh |
| Feed/magazine capacity: | 5-round rotary or 10-round box magazine |
| Operation: | Bolt action |
| Muzzle velocity: | 860mps (2820fps) |
| Effective range: | 1000m (3250ft) |

# Infantry Rifle Model 1889

One of Fabrique Nationale's earliest jobs was manufacturing an adaptation of the German Mauser rifle for the Belgian market. Mauser had just introduced the clip-loading method of magazine replenishment (in response to Mannlicher developments) and this was carried through to the Belgian weapon for the new 7.65 x 53mm cartridge specific to the Belgian gun. Several features separated the Belgian Mauser from its German equivalent. The bolt handle was set at the back of the receiver bridge and the barrel encased in a thin steel jacket, the intention being to detach it from the distortions of the wooden furniture. The gun was also cocked on the closing of the bolt, rather than the opening, although Mauser changed this configuration for subsequent guns, as it slowed the speed of bolt operation and thus the rate of fire.

| | |
|---|---|
| Country of origin: | Belgium |
| Calibre: | 7.65 x 53mm Belgian Mauser |
| Length: | 1295mm (50.98in) |
| Weight: | 4.01kg (8.82lb) |
| Barrel: | 780mm (30.6in), 4 grooves, rh |
| Feed/magazine capacity: | 5-round integral box magazine |
| Operation: | Bolt action |
| Muzzle velocity: | 610mps (2000fps) |
| Effective range: | 1000m (3250ft) |

# Artillery Musketoon Mle 1892

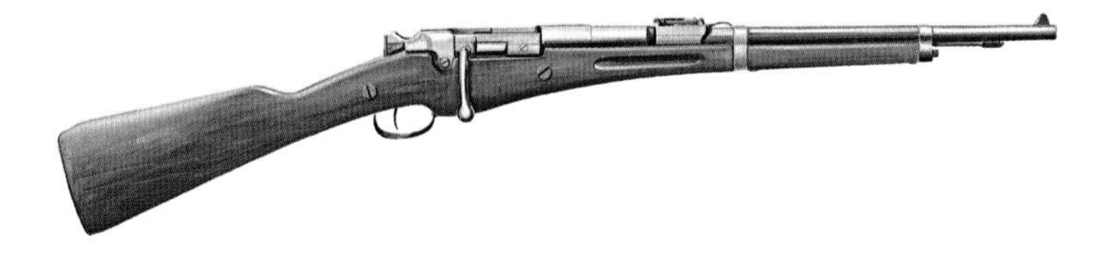

The Mle 1892 was one of the more distinctive-looking rifles developed in the late 19th century by André Berthier. It was part of a series of carbines (labelled 'Musketoon') which used the Mannlicher system of loading via a charger of rounds pushed down through the bolt into an integral magazine. Until 1915 (after some hard lessons were learned in the early stages of World War I), the three-round clip was used, then a new rifle was brought out which used a five-round clip. The Mle 1892 was a short, stocky rifle with a bulbous stock just in front of the trigger for the magazine housing and a cleaning rod running down the forestock. The short barrel and 8mm Lebel round only gave it a range of about 500m (1625ft), but this was more than adequate for most combat roles – most contemporary cartridges were overpowerful for the role for which they were required.

| | |
|---|---|
| **Country of origin:** | France |
| **Calibre:** | 8 x 50R Belgian Mauser |
| **Length:** | 940mm (37in) |
| **Weight:** | 3.1kg (6.8lb) |
| **Barrel:** | 445mm (17.5in), 4 grooves, lh |
| **Feed/magazine capacity:** | 3-round integral box magazine, clip-loaded |
| **Operation:** | Bolt action |
| **Muzzle velocity:** | 610mps (2000fps) |
| **Effective range:** | 500m (1640ft) |

# Lebel M1886

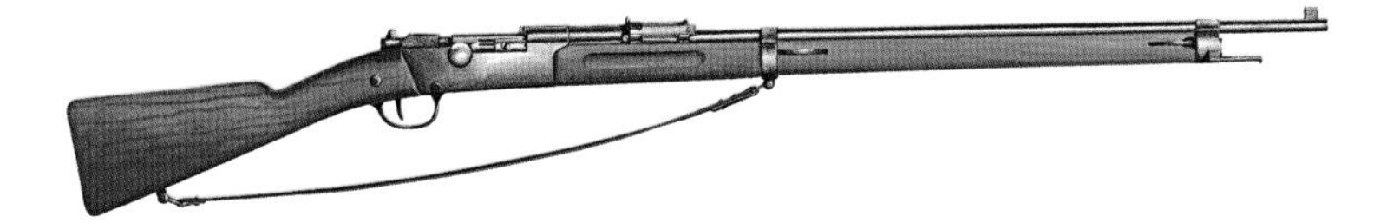

Although the M1886 first went into production in the late 1800s, its place in the history of 20th century small arms is confirmed by its being the standard French rifle of World War I. In effect, the rifle was not a new one, simply a 1874 Gras rifle updated for use with an eight-round tubular magazine which was fitted beneath the barrel. This, and the fact that the M1886 was the first rifle to use smokeless powder in conjunction with an 8mm bullet, made it a truly advanced weapon for its time. It was, however, a heavy gun to use – empty weight alone came to more than 4kg (8.8lb). Furthermore, magazine loading could be a long and awkward business (although single rounds could also be loaded directly into the chamber) and the bolt action was susceptible to dirt. Despite these flaws, the Lebel carried the French forces throughout World War I.

| | |
|---|---|
| Country of origin: | France |
| Calibre: | 8 x 50R Lebel Mle 1886 |
| Length: | 1295mm (50.98in) |
| Weight: | 4.28kg (9.44lb) |
| Barrel: | 800mm (31.5in), 4 grooves, lh |
| Feed/magazine capacity: | 8-round under-barrel tube magazine |
| Operation: | Bolt action |
| Muzzle velocity: | 715mps (2346fps) |
| Effective range: | 1000m (3250ft) |

# Lebel-Berthier 1907/15

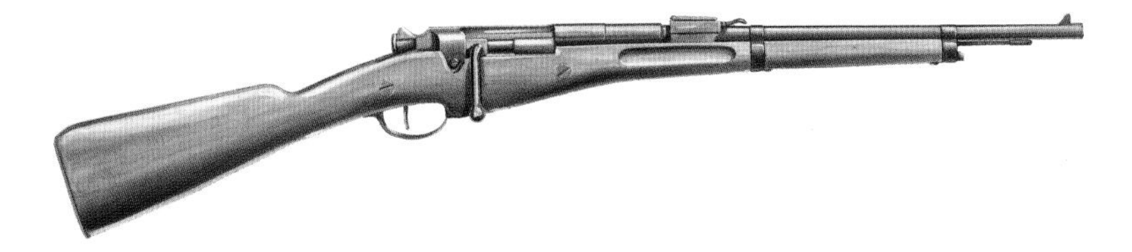

**During the late 1800s, the French military authorities commissioned a new rifle to be designed in response to German and Austro-Hungarian weapon improvements. The result was a long series of carbines which issued from a committee headed by the estimable André Berthier. The first weapon was the Mousqueton Berthier Mle 1890, followed by the Mle 1892. Both were short, portable weapons which used the same bolt-action system as the Lebel 1886, although a violent recoil and muzzle-flash detracted from their initial popularity with front-line troops. The Mle 1907/15 was actually produced as a replacement for the standard Lebel rifle and was a slightly modified Mle 1907 Senegal rifle (known as the 'Colonial' model). It even went on to contract manufacture in the USA under Remington Arms-Union Cartridge Company.**

| | |
|---|---|
| Country of origin: | France |
| Calibre: | 8 x 50R Lebel |
| Length: | 1303mm (51.3in) |
| Weight: | 3.79kg (8.35lb) |
| Barrel: | 798mm (31.42in), 4 grooves, lh |
| Feed/magazine capacity: | 3-round integral box magazine |
| Operation: | Bolt action |
| Muzzle velocity: | 715mps (2345fps) |
| Effective range: | 1000m (3250ft) |

# FR-F1

The FR-F1 was a French sniper rifle developed in the 1960s and based on the MAS 36 service rifle. The redesign was extensive: the barrel was lengthened, the stock was given a cheek rest, the trigger unit acquired a pistol grip and the barrel a flash hider. All these improvements produced a sniper rifle of competition standard, yet there were some deficiencies. Principal amongst these was the retention of the ageing 7.5mm Mle 1929 French Service round – the rest of Europe was modelling its comparable weapons around the now-standard 7.62mm NATO calibre ammunition. Eventually, later models of the FR-F1 were made in the NATO calibre, and this was continued in the FR-F2 model, which replaced the FR-F1. Other problems with the FR-F1 included a rather excessive weight for its role and a somewhat insubstantial bipod.

| | |
|---|---|
| Country of origin: | France |
| Calibre: | 7.5 x 54 mle 1929 and 7.62 x 51mm NATO |
| Length: | 1138mm (44.8in) |
| Weight: | 5.2kg (11.46lb) |
| Barrel: | 552mm (21.37in), 4 grooves, rh |
| Feed/magazine capacity: | 10-round integral box magazine |
| Operation: | Bolt action |
| Muzzle velocity: | 852mps (2795fps) |
| Effective range: | 800m (2624ft) |

# Mauser SP66

The SP66 is one of Mauser's top-flight bolt-action sniper rifles, a heavily machined and prodigiously accurate weapon modified from a civilian hunting rifle. Sophistication runs through the weapon from muzzle to butt. The gun uses Mauser's short-action bolt operation, which allows the user to keep his head in place (and thus observing the target) while reloading. The stock is fully adjustable and has a thumb hold for extra grip control, and surfaces are deliberately roughened to aid overall hand adhesion. Sighting is done through a x1.5 – x6 Zeiss-Divari telescopic sight (the gun can also accept almost any other sight commercially available) and the muzzle is fitted with a combined muzzle brake and flash reducer. The Mauser's only drawback is that the quality of the SP66 makes it an expensive weapon; Mauser make the 86SR as a less costly version.

| | |
|---|---|
| Country of origin: | Germany |
| Calibre: | 7.62mm NATO |
| Length: | 1210mm (47.64in) |
| Weight: | 6.12kg (13.5lb) with telescopic sight |
| Barrel: | 650mm (25.59in); 750mm (28.74in) with muzzle brake; 4 grooves, rh |
| Feed/magazine capacity: | 3-round integral box magazine |
| Operation: | Bolt action |
| Muzzle velocity: | 868mps (2848fps) |
| Effective range: | 1000m (3250ft) |

# Mauser Gewehr 98

The Mauser Gewehr 98 was in production from 1898–1918 and was the standard German rifle during World War I. It is the archetypal Mauser weapon, and the bolt action established a model which is followed to this day. The bolt action consisted of three locking lugs, the third being beneath the bolt and which dropped into a recess in the receiver for extra locking safety. The gun was charger-loaded with no visible magazine and it proved itself a solid, reliable and accurate rifle which served soldiers well. It was consistently well crafted, although there was some deterioration in quality as the war ground on. Like most of its contemporaries, it was much more powerful than it needed to be for the time. World War II saw the Gewehr 98 in action once more, although by then it was mostly being replaced by a shortened version, the KAR 98.

| | |
|---|---|
| Country of origin: | Germany |
| Calibre: | 7.92mm Mauser M98 |
| Length: | 1255mm (49.4in) |
| Weight: | 4.14kg (9.13lb) |
| Barrel: | 740mm (29.14in), 4 grooves, rh |
| Feed/magazine capacity: | 5-round integral box magazine |
| Operation: | Bolt action |
| Muzzle velocity: | 870mps (2855fps) |
| Effective range: | 1000m (3250ft) plus |

# Mauser Kar 98K

German experiments with a carbine began, like many other European nations, in the late 1890s. The first issue was the M1898, otherwise known as the Kar 98, which was produced between 1899 and 1903, when it was revised to make the M1904. This latter weapon became the standard German rifle of World War I and, in 1920, was labelled as the Kar 98a. This was still felt to be too long for convenient infantry use and by 1939 it had been further shortened to produce the Karabiner 98k (Kar 98k). In fact, there was little to separate the 98k from the 98a. Visually it could be distinguished by its recess in the forestock and a less exposed muzzle, but what really separated the 98k was its prevalence – it became the standard German rifle of World War II. Even though quality of materials deteriorated by the end of the war, the 11.5 million Kar 98ks produced gave reliable service well after 1945.

| | |
|---|---|
| **Country of origin:** | Germany |
| **Calibre:** | 7.92mm Mauser M98 |
| **Length:** | 1110mm (43.7in) |
| **Weight:** | 3.9kg (8.6lb) |
| **Barrel:** | 600mm (23.62in), 4 grooves, rh |
| **Feed/magazine capacity:** | 5-round integral box magazine |
| **Operation:** | Bolt action |
| **Muzzle velocity:** | 745mps (2444fps) |
| **Effective range:** | 1000m (3250ft) plus |

# Lee-Enfield Mk II

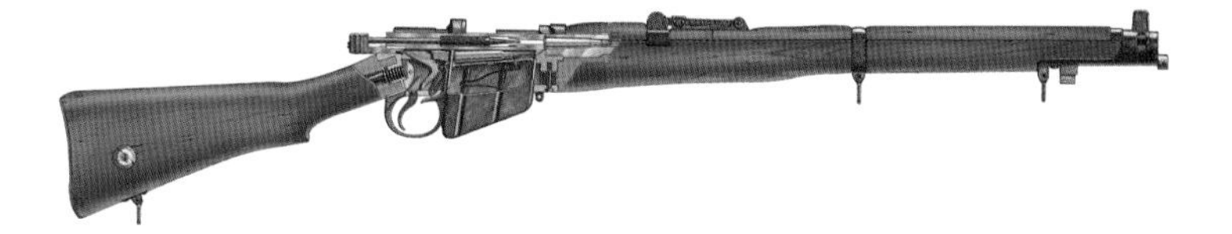

The Lee-Enfield rifles are some of the best designed bolt-action rifles in history. They actually emerged out of a fusion between the smooth bolt operation of the earlier Lee-Metford rifles (so called after the designers James Paris Lee and William Metford) with a rifling developed at the Royal Small Arms Factory, Enfield Lock. The first gun produced was the Lee-Enfield Mk I, issued from 1895, which had the new rifling and altered sights, and saw service in South Africa. In 1903, there arrived the Short Magazine Lee-Enfield Mk I. This began the true 'SMLE' configuration which would become the standard British infantry rifle format for many years. It had 125mm (5in) removed from its barrel to reduce its length. The Mk II was not radically different from the Mk I, except that it had new sights and barrel (the latter being shorter and lighter) and could be loaded by chargers.

| | |
|---|---|
| Country of origin: | Great Britain |
| Calibre: | .303in British Service |
| Length: | 1132mm (44.57in) |
| Weight: | 3.71kg (8.18lb) |
| Barrel: | 640mm (25.19in), 5 grooves, lh |
| Feed/magazine capacity: | 10-round detachable box magazine |
| Operation: | Bolt action |
| Muzzle velocity: | 617mps (2025fps) |
| Effective range: | 1000m (3250ft) plus |

# Short Magazine Lee-Enfield Mk III

**From its entry into service in 1907, the Short Magazine Lee- Enfield (SMLE) Mk III became one of the seminal small arms of the 20th century. Officially designated Rifle No. 1, Mk III, it used the bolt action developed by the US arms designer James Lee and was produced at the Royal Small Arms Factory at Enfield Lock, United Kingdom, hence its name. Its virtues were of supreme value to British and Commonwealth soldiers across both world wars: a smooth, fast bolt action which could give 15rpm in trained hands; its capacious magazine which held 10 rounds and single rounds which could be directly loaded even with a full magazine; a sighted range of around 1000m (3280ft); rugged enough to withstand combat duties. Such was the quality of this gun that some of the three million produced still crop up in combat across the world.**

| | |
|---|---|
| **Country of origin:** | Great Britain |
| **Calibre:** | .303in British Service |
| **Length:** | 1133mm (44.6in) |
| **Weight:** | 3.93kg (8.65lb) |
| **Barrel:** | 640mm (25.2in), 5 grooves, lh |
| **Feed/magazine capacity:** | 10-round detachable box magazine |
| **Operation:** | Bolt action |
| **Muzzle velocity:** | 670mps (2300fps) |
| **Effective range:** | 1000m (3250ft) plus |

# Lee-Enfield Rifle No. 4 Mk 1

**The beginning of World War II saw the SMLE Mk III as the standard British Army rifle, yet, by 1940, the need for a new version more amenable to wartime production restraints became evident. The Lee-Enfield Rifle No. 4 Mk 1 was the result – a weapon that, along with the Sten submachine gun and Bren machine gun, became synonymous with British troops. Visually, the most obvious change was the muzzle; the flat muzzle nosecap was cut back about 5cm (2in), with the exposed barrel taking direct fitment of the foresight and the new spike bayonet that replaced the sword bayonet. The rear sights were also moved to above the end of the bolt. A Mk 1* with a simplified receiver followed shortly after the introduction of the Mk 1. Like the SMLE, the No. 4 rifle was reliable, durable and had searching accuracy. About four million went through production.**

| | |
|---|---|
| **Country of origin:** | Great Britain |
| **Calibre:** | .303in British Service |
| **Length:** | 1128mm (44.43in) |
| **Weight:** | 4.11kg (9.06lb) |
| **Barrel:** | 640mm (25.2in), 5 grooves, lh |
| **Feed/magazine capacity:** | 10-round detachable box magazine |
| **Operation:** | Bolt action |
| **Muzzle velocity:** | 751mps (2464fps) |
| **Effective range:** | 1000m (3250ft) plus |

# Lee-Enfield Rifle No. 5

Lee-Enfield's Rifle No. 5 was the least satisfying of its bolt-action guns. It was born out of the demand from 1943 for a more compact weapon for use in the British Army's jungle campaigns in the Far East. The length of the SMLE in the jungle proved awkward, and an adaptation of the trusty Lee Enfield seemed the best option. The No. 5 was basically a No. 4 Rifle with a shortened forestock and barrel, the former featuring a rubber shoulder protector and the latter having a large flash hider. Both were essential, as the .303 round gave huge blast and recoil when fired from the new barrel length, but both modifications were ultimately inadequate for controlling the cartridge's power. This, combined with a flaw in the gun's sighting which meant that zero would shift from one day to the next, led to the gun being abandoned in 1947.

| | |
|---|---|
| Country of origin: | Great Britain |
| Calibre: | .303in British Service |
| Length: | 1000mm (39.37in) |
| Weight: | 3.24kg (7.14lb) |
| Barrel: | 478mm (18.7in), 5 grooves, lh |
| Feed/magazine capacity: | 10-round detachable box magazine |
| Operation: | Bolt action |
| Muzzle velocity: | 610mps (2000fps) |
| Effective range: | 1000m (3250ft) |

# De Lisle Carbine

Many silenced weapons have entered the market since the De Lisle's birth in World War II, but few have surpassed the sheer level of noise reduction it attained. Its exceptional suppressor eliminates the muzzle report almost entirely, the only noises left being the bolt operation and the striker hitting the cap. Using a .45 ACP round (its magazine was that of the Colt M1911A1 pistol), it was still able to hit a target up to 400m (1312ft) away both accurately and powerfully. The British Commandos were the De Lisle's original users during the war, but, in the postwar period, many other special forces units utilised it when the need for accurate, silent killing arose. Indeed, some companies were still producing the De Lisle to order in the 1980s. Its only potential drawback was the need to work the bolt between each shot, which could expose the firer through his or her movement.

| | |
|---|---|
| Country of origin: | Great Britain |
| Calibre: | .45 ACP |
| Length: | 960mm (37.79in) |
| Weight: | 3.7kg (8.15lb) |
| Barrel: | 210mm (8.26in), 4 grooves, lh |
| Feed/magazine capacity: | 7-round detachable box magazine |
| Operation: | Bolt action |
| Muzzle velocity: | 260mps (853fps) |
| Effective r ange: | 400m (1312ft) |

# Parker-Hale Model 85

The Model 85 is one of a series of fine sniper and hunting weapons to emerge from the factories of Parker-Hale in recent years (although the manufacturer since 1990 is actually the Gibbs Rifle Co., USA). As a modern bolt-action sniper rifle, it is fairly standard; it has an adjustable butt and cheek rest, a detachable bipod and fires the 7.62mm NATO round. Yet the quality of its production, particularly in its heavy chrome-molybdenum free-floating barrel and Mauser-style bolt action, makes it very accurate indeed. The standard x6 telescopic sight can give a 100 per cent first-round hit rate at up to 600m (1968ft) and an 85 per cent hit rate at up to 900m (2952ft) when used by a skilled professional. Despite such sterling accuracy, the British Army, for whom the Model 85 was designed, rejected it in favour of the Accuracy International L96A1.

| | |
|---|---|
| Country of origin: | Great Britain |
| Calibre: | 7.62 x 51mm NATO |
| Length: | 1150mm (45.28in) |
| Weight: | 5.7kg (12.57lb) with telescopic sight |
| Barrel: | 700mm (27.56in), 4 grooves, rh |
| Feed/magazine capacity: | 10-round box magazine |
| Operation: | Bolt action |
| Muzzle velocity: | 860mps (2820fps) |
| Effective range: | 1000m (3250ft) plus |

# RSAF L42A1

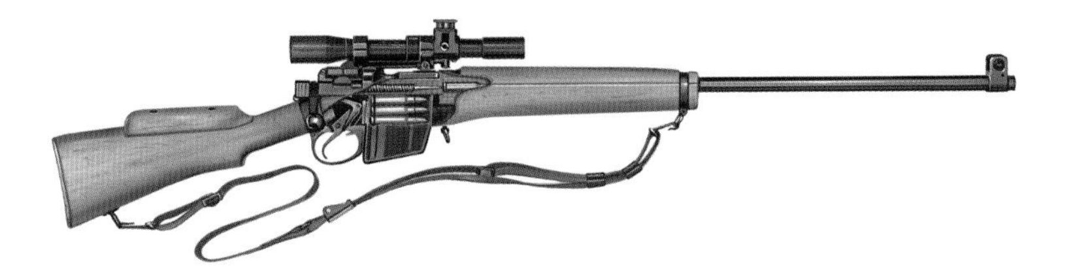

Such is the essential credibility of the Lee-Enfield rifles as high-accuracy weapons that a version is still used today by British Army and Royal Marine snipers, although it is now being phased out. The L42A1 is a 7.62mm conversion of the old .303 calibre No. 4 Mk 1 or Mark 1*(T) rifle, using the old L1 telescopic sight, a sight which in trained hands and with good ammunition (usually 'Green Spot' cartridges made by Royal Ordnance) gives the rifle consistent first-shot accuracy up to 1000m (3280ft). Although the Lee-Enfield heritage of the rifle is visually clear, new modifications in the weapon include a different barrel, a 10-round magazine adapted to the rimless 7.62mm round, a shortened forestock and a cheek rest on the butt. Civilian and police versions are also produced, respectively known as Envoy and Enforcer.

| | |
|---|---|
| Country of origin: | Great Britain |
| Calibre: | 7.62 x 51mm NATO |
| Length: | 1181mm (46.5in) |
| Weight: | 4.43kg (9.76lb) |
| Barrel: | 699mm (27.5in), 4 grooves, rh |
| Feed/magazine capacity: | 10-round detachable box magazine |
| Operation: | Bolt action |
| Muzzle velocity: | 838mps (2750fps) |
| Effective range: | 1000m (3280ft) plus |

# Accuracy International L96A1

The L96A1 entered production in 1985 as the replacement for the Lee-Enfield L42A1 as the British Army and Royal Marine's standard sniper rifle. It continues in 7.62mm NATO calibre, but, being descended from the Model PM, an Olympic-standard sports rifle, it has a new level of sophistication for a combat sniper weapon. The free-floating stainless steel barrel and Tasco sight give a 100 per cent hit rate at 600m (1968ft), while the stock is designed for ambidextrous use. The rifle is fitted with an alloy bipod and both rifle stripping and barrel change are simple tasks. Variants of the L96A1 include versions chambered for 7mm Remington Magnum, .300 Winchester Magnum and .338 Lapua Magnum (rounds which give extra range), a single-shot long-range version and a silenced version which uses subsonic ammunition.

| | |
|---|---|
| Country of origin: | Great Britain |
| Calibre: | 7.62mm NATO and others |
| Length: | 1163mm (45in) |
| Weight: | 6.2kg (13.68lb) |
| Barrel: | 654mm (26in), 4 grooves, rh |
| Feed/magazine capacity: | 10-round detachable box magazine |
| Operation: | Bolt action |
| Muzzle velocity: | 840mps (2830fps) |
| Effective range: | 1000m (3250ft) |

# L115A3 / AWM

First deployed in 2008, the L115A3 is used by snipers in the British Army, Royal Marines and the RAF Regiment. The world record for the longest sniper kill – actually two kills in rapid succession – was established by British corporal Craig Harrison at 2475m (8119ft) using the L115A3 against the Taliban in Helmand Province, Afghanistan, in 2009. Manufactured by Accuracy International, the L115A3 AWM – which stands for Arctic Warfare Magnum – is a development of the .338 Lapua Magnum L115A1 Long Range Rifle. The L115A3 features include Schmidt & Bender 5-25x56 telescopic sights, suppressors to reduce the flash and noise signature, adjustable cheek pieces for better eye alignment with the telescopic sights, and adjustable bipods.

| | |
|---|---|
| Country of origin: | Great Britain |
| Calibre: | 7.62mm (.3in) / .300 Winchester Magnum, 8.58mm (.338in) / .338 Lapua Magnum |
| Length: | 1300mm (51in) |
| Weight: | 6.8kg (15lb) |
| Barrel: | 686mm (27in) |
| Feed/magazine capacity: | 5-round detachable box magazine |
| Operation: | Bolt action |
| Muzzle velocity: | c.850m/sec (2788ft/sec) |
| Effective range: | 1100m (3609ft) .300 Winchester; 1500m (4921ft) .338 |

# Mannlicher-Carcano 1891

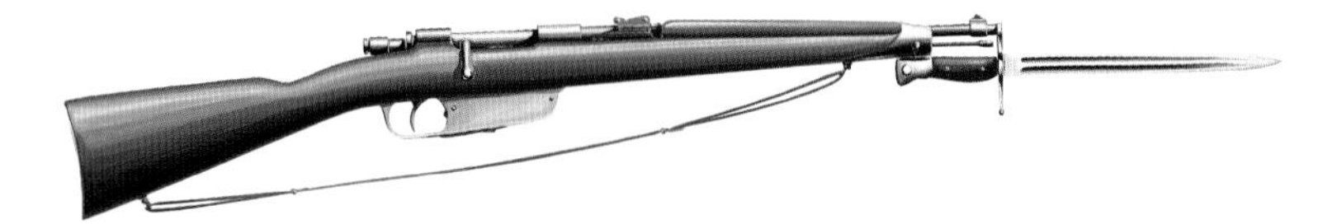

The Mannlicher-Carcano rifle began its life in the 1890s in Italy and established a pattern of rifles that would equip many Italian soldiers throughout both world wars. Its name comes from a combination of that of Salvatore Carcano, the rifle's overall designer, and that of the Mannlicher weapons, although only the Mannlicher magazine type was actually adopted in the final gun which emerged at the end of the design process. The first Mannlicher-Carcano issue was the M1891. This was a 6.5mm weapon which used a Mauser bolt action and a six-round integral box magazine that was loaded by chargers. The M1891 was a solid enough gun which became the standard Italian army weapon during World War I; there is little remarkable about its design or its performance. It did, however, set the scene for an interesting series of carbine models.

| | |
|---|---|
| Country of origin: | Italy |
| Calibre: | 6.5 x 52mm Mannlicher Carcano |
| Length: | 1290mm (50.79in) |
| Weight: | 3.8kg (8.38lb) |
| Barrel: | 780mm (30.6in), 4 grooves, rh |
| Feed/magazine capacity: | 6-round integral box magazine |
| Operation: | Bolt action |
| Muzzle velocity: | 730mps (2400fps) |
| Effective range: | 1000m (3250ft) plus |

# Moschetto Modello 1891 per Cavalleria

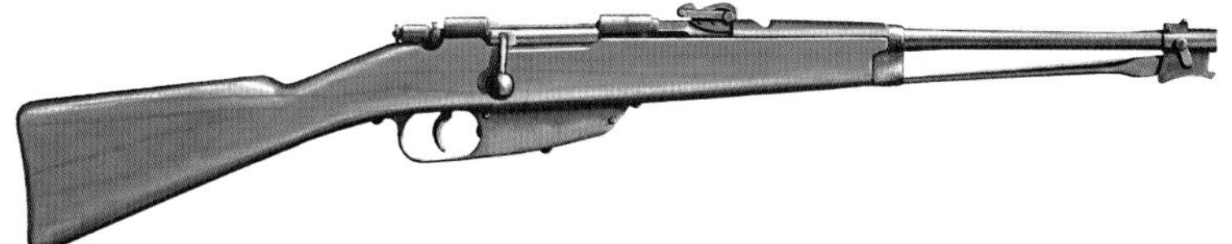

Like many European rifles in the 1890s, the Mannlicher-Carcano guns were soon recognised to be in need of a shortened, carbine version for use by cavalry troops. This need produced the Moschetto Modello 1891 per Cavalleria, otherwise known as the 'Truppo Speciale' model. It retained the same operating features as the parent rifle, but the length was quite dramatically reduced to 920mm (36.2in). Other features included a fixed bayonet which could be folded back when not in use to lie just beneath the barrel. Although generally intended for cavalry use, it also became a popular rifle with many auxiliary units, such as artillerymen, engineers and signallers, whose work demanded that weapons had compact dimensions for storage. Yet the size advantage could not alleviate the problems created in combat by the M91 series' underpowered 6.5mm cartridge.

| | |
|---|---|
| Country of origin: | Italy |
| Calibre: | 6.5 x 52mm Mannlicher Carcano |
| Length: | 920mm (36.2in) |
| Weight: | 3kg (6.62lb) |
| Barrel: | 610mm (24in) |
| Feed/magazine capacity: | 6-round integral box magazine |
| Operation: | Bolt action |
| Muzzle velocity: | 700mps (2275fps) |
| Effective range: | 600m (1950ft) |

# Beretta Sniper

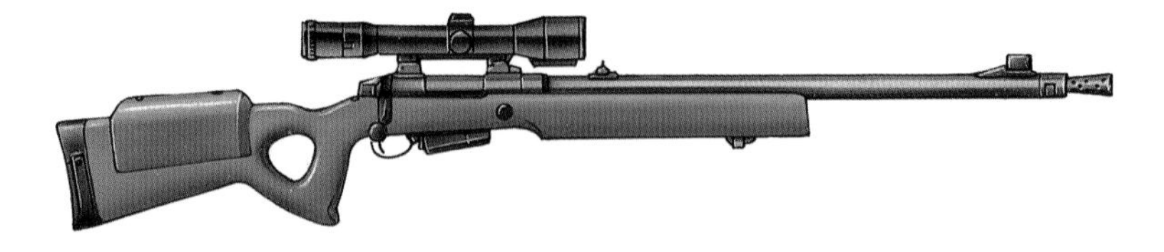

Despite being a rather conventional rifle, the Beretta Sniper still possesses a very good performance, due in the main to Beretta's superb standards of workmanship and engineering. What it does have, however, is an advanced free-floating barrel design, in which a counterweight in the forestock serves almost to cancel out barrel vibrations upon firing for even greater accuracy. Generally used in a sniper role with the standard-issue Zeiss Divari Z telescopic sights, it can also be fitted with precision iron sights which are capable of telling use in trained hands. The Beretta Sniper has not achieved great sales outside of Italian security use and appears to be a little undersold. Yet it is still a fine weapon compatible with standard NATO ammunition, so could make a more overt appearance in the world market in the future.

| | |
|---|---|
| Country of origin: | Italy |
| Calibre: | 7.62 x 51mm NATO |
| Length: | 1165mm (45.87in) |
| Weight: | 5.55kg (12.23lb) |
| Barrel: | 586mm (23.07in), 4 grooves, rh |
| Feed/magazine capacity: | 5-round detachable box magazine |
| Operation: | Bolt action |
| Muzzle velocity: | 840mps (2755fps) |
| Effective range: | 1000m (3250ft) plus |

# Arisaka 38th Year Rifle

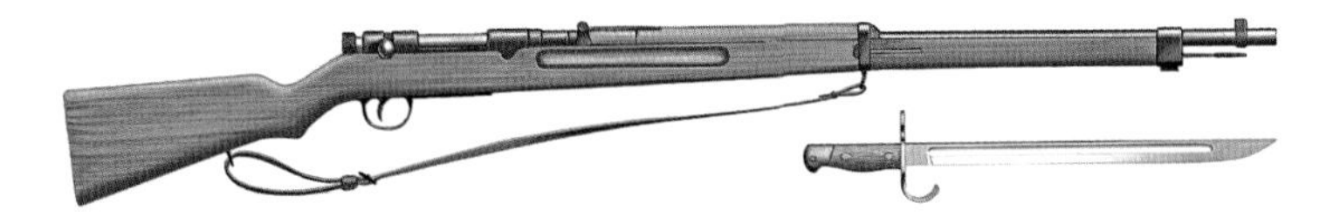

The Arisaka 38th Year Rifle took its name from its designer, Colonel Arisaka, and the year of the Japanese Emperor's reign. It became the standard Japanese infantry rifle between 1907 and 1944, and comprised a mix of Mauser and Mannlicher features which had already produced the 30th Year Rifle in the late 1890s. In many ways, the 38th Year was a success because it was built with the Japanese frame in mind. Its 6.5mm cartridge was low-powered enough to be controlled by a smaller figure, while the pronounced length of the rifle gave some advantage in bayonet clashes against longer-limbed Caucasian opponents. A sound enough weapon, the 38th Year Rifle gave good service until the last years of World War II, when shortages of raw materials resulted in poorly constructed and sometimes dangerously unstable weapons.

| | |
|---|---|
| Country of origin: | Japan |
| Calibre: | 6.5mm Japanese Service |
| Length: | 1275mm (50.25in) |
| Weight: | 4.31kg (9.5lb) |
| Barrel: | 798mm (31.45in), 6 grooves, rh |
| Feed/magazine capacity: | 5-round internal box magazine |
| Operation: | Bolt action |
| Muzzle velocity: | 730mps (2400fps) |
| Effective range: | 1000m (3250ft) plus |

# Springfield Model 1903

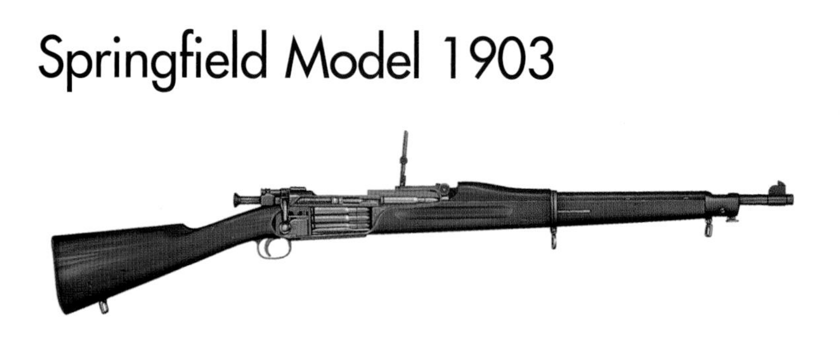

**Despite taking the name of the Springfield arsenal in Illinois, the Model 1903 actually originated with Mauser, who were asked to develop a new US service rifle under licence in the USA to replace the Krag-Jorgeson rifle used by US Army servicemen since 1892. The gun was first produced around the flat-ended .30 M1903 cartridge, then around the pointed M1906 (same calibre). From 1903, however, it was in production at Springfield and would stay in production until 1965. The reasons for its longevity were simple. With dimensions between full rifle and carbine, it was convenient to carry; it was accurate enough for sniper use; it rarely failed; and it was comfortable to use. A run of variations followed the original model (which can be distinguished by its straight stock, rather than pistol grip), including the M1903A4 sniper gun used in Vietnam.**

| | |
|---|---|
| Country of origin: | USA |
| Calibre: | .30in M1906 |
| Length: | 1097mm (43.19in) |
| Weight: | 3.94kg (8.68lb) |
| Barrel: | 610mm (24in), 4 grooves, rh |
| Feed/magazine capacity: | 5-round internal box magazine |
| Operation: | Bolt action |
| Muzzle velocity: | 853mps (2800fps) |
| Effective range: | 1000m (3250ft) plus |

# Enfield Rifle M1917

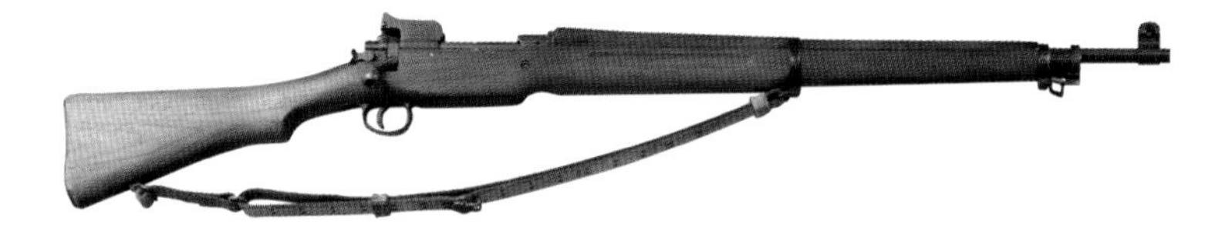

**The aborted British development of the .303in British Enfield Rifle No. 2 was not to be wasted on the USA, who rechambered the weapon successfully for the .30-06 rimless US Service cartridge in 1917. A shortage of rifles at the beginning of the US part of World War I was behind the redevelopment and more than two million of the M1917 were subsequently made in two years of production before the end of the war in November 1918. The British and US weapons were almost identical in every way. Indeed, when more than 100,000 M1917s were supplied to the British Home Guard during World War II, a red stripe had to be painted around the stock to stop soldiers accidentally loading the rifles with the rimmed .303 British Service round. The M1917 also served after World War I in the hands of commercial buyers in the USA.**

| | |
|---|---|
| Country of origin: | USA |
| Calibre: | .30-06 US Service |
| Length: | 1174mm (46.25in) |
| Weight: | 4.36kg (9.61lb) |
| Barrel: | 660mm (26in), 5 grooves, lh |
| Feed/magazine capacity: | 5-round detachable box magazine |
| Operation: | Bolt action |
| Muzzle velocity: | 853mps (2800fps) |
| Effective range: | 1000m (3250ft) plus |

# Weatherby Mk V

The Weatherby Mk V is actually a sporting rifle used for hunting, not a military gun, but its capabilities are just as impressive and its overall quality of build and design is excellent. Its US designer was one Roy Weatherby and the Mk V designation is actually the name of the locking system used in the bolt action. This used some nine locking lugs configured to give the exceptionally smooth action which hunters appreciate when needing a rapid and solid reloading. As a commercial rifle, the Mk V has been through a bewildering variety of calibres to suit different pursuits and hunting limitations. These have ranged from .22-250 to the potent .460 Winchester Magnum calibre, and the consequent change of barrels means that there is much variety in the weight, velocity and effective range of Weatherby guns.

| | |
|---|---|
| Country of origin: | USA |
| Calibre: | Various |
| Length: | 1105mm (43.5in) to 1180mm (46.5in) |
| Weight: | 2.95kg (6.5lb) to 4.75kg (10.5lb) |
| Barrel: | 610mm (24in) or 660mm (26in) |
| Feed/magazine capacity: | 5-round integral box magazine |
| Operation: | Bolt action |
| Muzzle velocity: | Various |
| Effective range: | 1000m (3250ft) plus |

# M40A1

The M40 was a specific selection of the US Marine Corps in 1966 and is essentially a standard Remington Model 700 adapted for military use. The M40 was a superb weapon by all accounts and authorities, and met the high standards placed on it by the exhaustively trained Marine snipers. It featured a Mauser-type bolt action and a heavy barrel, with a five-round magazine. The M40A1 is an improved version of this gun. The heavy barrel has given way to a stainless steel barrel and the furniture is now made out of the lighter fibreglass, rather than wood, which improves its handling. In addition, whereas the M40 used a Redfield zoom telescopic sight which could reach x9 magnification, the latest sight climbs to x10 magnification. This sight obviates the need for the iron sight fitting which can be seen on some of the earliest M40 models.

| | |
|---|---|
| Country of origin: | USA |
| Calibre: | 7.62 x 51mm NATO |
| Length: | 1117mm (43.98in) |
| Weight: | 6.57kg (14.48lb) |
| Barrel: | 610mm (24in), 4 grooves, rh |
| Feed/magazine capacity: | 5-round integral box magazine |
| Operation: | Bolt action |
| Muzzle velocity: | 777mps (2550fps) |
| Effective range: | 800m (2624ft) plus |

# McMillan TAC-50

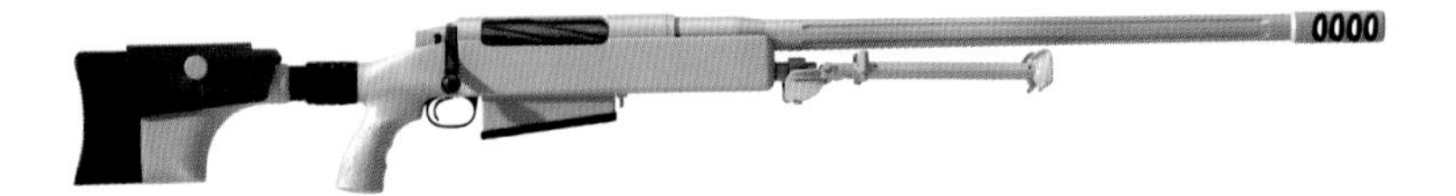


**Firing extremely powerful ammunition adapted from a heavy machine gun round, the TAC-50 was used to make what was then the world's longest confirmed kill in Afghanistan in March 2002. In the same month Master Corporal Arron Perry killed an enemy combatant from 2310m (7579ft) and Corporal Rob Furlong killed an enemy combatant at a distance of 2430m (7972ft). The gun's stock is made from fibreglass and is designed to be used from a bipod only. A manually operated rifle, the TAC-50 has no open sights and can be used with a variety of telescopic or night sights. The large bolt has dual front locking lugs, and its body has spiral flutes to reduce weight. The TAC-50 is in service with, among other users, French Navy commandos, Israeli special forces and the US Navy SEALs, where it is designated the Mk 15.**

| | |
|---|---|
| Country of origin: | USA |
| Calibre: | 12.7mm (.5in) |
| Length: | 1448mm (57in) |
| Weight: | 11.8kg (26lb) |
| Barrel: | 736mm (29in) |
| Feed/magazine capacity: | 5-round detachable box |
| Operation: | Manually-operated rotary bolt action |
| Muzzle velocity: | 823mps (2700fps) |
| Effective range: | 1600m (5249ft) |

# Mosin-Nagant Rifle

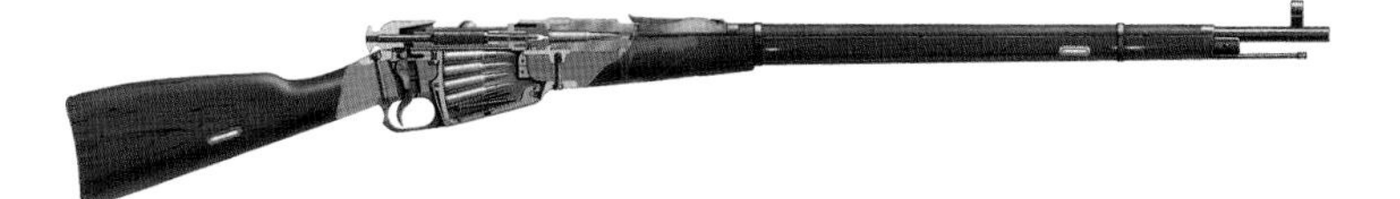

The Mosin-Nagant Rifle was a mix of designs from the Belgian brothers Emil and Leon Nagant and the Tsarist Russian officer Sergei Mosin. It was produced as a new standard rifle for Russian infantry forces and was issued as the M1891. Like most Russian weapons throughout the century, the key property of the Mosin-Nagant was its durability under the worst conditions. This was somewhat surprising, as the rifle had some fairly complex design features. Foremost amongst these was the two-piece bolt action, which actually allowed the cartridge being fired to remain free from magazine spring pressure. Yet the gun worked well and it became the standard Russian and Soviet service rifle (although hopelessly under-produced to meet demand) and emerged in several carbine models (the World War II model was the M1891/30).

| | |
|---|---|
| **Country of origin:** | Russia/USSR |
| **Calibre:** | 7.62 x 54R Mosin-Nagant |
| **Length:** | 1304mm (51.25in) |
| **Weight:** | 4.43kg (9.77lb) |
| **Barrel:** | 802mm (31.6in) |
| **Feed/magazine capacity:** | 5-round integral box magazine |
| **Operation:** | Bolt action |
| **Muzzle velocity:** | 805mps (2650fps) |
| **Effective range:** | 1000m (3250ft) |

# Steyr-Mannlicher AUG

**The bullpup-design Steyr-Mannlicher Armee Universal Gewehr is one of the best assault rifles in existence today. Its rather fragile appearance masks an unswerving reliability under the hardest conditions, and it is very stable and accurate. Its weight is kept down by an extensive use of advanced plastics, even in areas such as the firing mechanism and its clear-plastic magazine, which allows the user to check ammunition usage. Fire selection is accomplished by a two-stage trigger and the firing mechanism is easily changed to vary the selective-fire configuration. Indeed, the whole rifle is modular in form and short and long barrels are interchangeable through a simple twist of the front grip. Standard in the Austrian Army, the AUG is spreading throughout the world in places such as Australia and the USA.**

| | |
|---|---|
| Country of origin: | Austria |
| Calibre: | 5.56mm M198 or NATO |
| Length: | 790mm (31.1in) |
| Weight: | 3.6kg (7.93lb) |
| Barrel: | 508mm (20in), 6 grooves, rh |
| Feed/magazine capacity: | 30- or 42-round detachable box magazine |
| Operation: | Gas |
| Cyclic rate of fire: | 650rpm |
| Muzzle velocity: | 970mps (3182fps) |
| Effective range: | 500m (1640ft) plus |

# FN FAL

The Fabrique Nationale Fusil Automatique Legère (FN FAL) rifle appeared in 1948 in 7.92mm calibre, soon changing to 7.62mm in accordance with NATO standardisation. More than 50 years later, it has found service in more than 90 countries across the world, gaining its popularity through fine machining, a long-range reach and solid reliability. The rifle works on a gas-operated system, the gas piston being situated in the housing above the barrel. This system is capable of full automatic fire – although the power of the cartridge meant that the weapon climbed in such a mode and most models were semi-automatic in form. One exception was the FN FAL (HB) (Heavy Barrel), which was fitted with a bipod as an infantry support weapon. The FN FAL's success spawned a myriad of variations seen the world over, including the L1A1, the standard British Army gun until the mid-1980s.

| | |
|---|---|
| Country of origin: | Belgium |
| Calibre: | 7.62mm NATO |
| Length: | 1053mm (41.46in) |
| Weight: | 4.31kg (9.5lb) |
| Barrel: | 533mm (21in), 4 grooves, rh |
| Feed/magazine capacity: | 20-round box magazine |
| Operation: | Gas, self-loading |
| Cyclic rate of fire: | 550rpm (full-automatic versions only) |
| Muzzle velocity: | 853mps (2800fps) |
| Effective range: | 800m (2600ft) plus |

# FN FAL Para

The FN FAL rifle comes in a multitude of forms and formats which vary depending on the country of use and the units for which it is destined. However, most standard FN FAL rifles are based around four key variants: the FN 50-00 (fixed stock, standard barrel); the FN 50-64 (side-folding stock, standard barrel); the FN 50-41 (fixed stock, heavy barrel, bipod); and the FN 50-63 Para model. This last weapon was designed to provide a more compact rifle for special forces and airborne use, and featured a folding skeletal stock and a shortened barrel. These new dimensions did not bring the FN FAL's overall performance down, but they did reduce its stock extended length to 770mm (30.3in) which meant that the rifle was more portable. The Para rifle, like the other FN FAL models, has achieved extensive world-wide distribution during its lifetime.

| | |
|---|---|
| Country of origin: | Belgium |
| Calibre: | 7.62 x 51mm NATO |
| Length: | 1020mm (40.15in) stock extended; 770mm (30.3in) stock folded |
| Weight: | 4.36kg (9.61lb) |
| Barrel: | 436mm (17.1in) 4 grooves, rh |
| Feed/magazine capacity: | 20-round detachable box magazine |
| Operation: | Gas operated |
| Cyclic rate of fire: | 650–700rpm |
| Muzzle velocity: | 853mps (2800) |
| Effective range: | 900m (2952ft) plus |

# FN FNC

The Fabrique Nationale FNC was based on the FN CAL (Carabine, Automatic, Legère), a 5.56mm update of the famous FN FAL rifle that had sold so well for the company. The FN CAL had been somewhat ahead of its time in calibre and features, and so was relatively unsuccessful, but the FN FNC was produced when the small-calibre, high-velocity rifle became more in vogue. The FNC found greater market success because, through the use of alloys, plastics, stampings and pressings, it was cheaper to manufacture than the CAL, and it could also meet with the ever-growing needs of standardisation by accepting the ubiquitous US M16 magazine. Use of the FNC has spread across the world since its appearance, particularly centred in countries such as Sweden (who designate it the AK5), Belgium and Indonesia (where it is manufactured under licence).

| | |
|---|---|
| Country of origin: | Belgium |
| Calibre: | 5.56 x 45mm NATO |
| Length: | 997mm (39.25in) stock extended; 766mm (30.15in) stock folded |
| Weight: | 3.8kg (8.38lb) |
| Barrel: | 449mm (17.68in), 6 grooves, rh |
| Feed/magazine capacity: | 30-round detachable box magazine |
| Operation: | Gas |
| Cyclic rate of fire: | 600–750rpm |
| Muzzle velocity: | 965mps (3165fps) |
| Effective range: | 500m (1640ft) plus |

# FN F2000

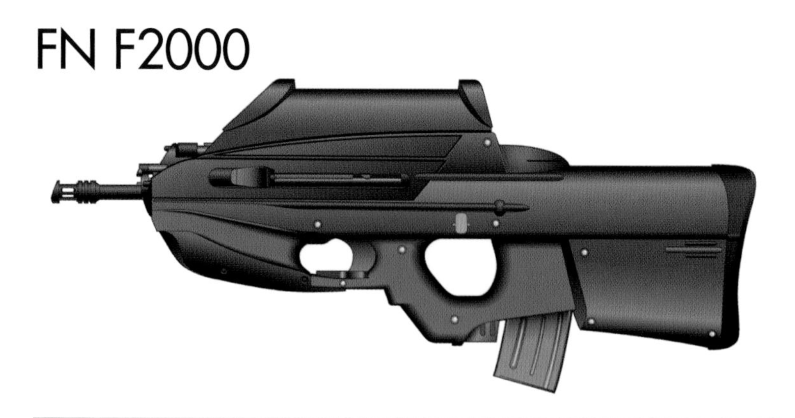

The FN F2000 is one of a new generation of assault rifles. Firing standard 5.56 x 45mm NATO rounds, it has a compact bullpup layout and a gas-operated, rotating-bolt (seven lugs) operating mechanism. One distinctive feature is the front ejection system; the spent cartridge cases are redirected from the chamber to an ejection port near the muzzle. This configuration aids accurate shooting (the shooter doesn't have to cope with hot cases ejecting close to his face) and makes the weapon ideally suited to firing through vehicle ports. Other virtues of the FN F2000 are its completely ambidextrous layout and its modularity; it can take numerous add-ons, from optical sights and laser rangefinders through to 40mm grenade launchers and riot-control weapons. It is used by the Belgian Special Forces Group, the Pakistani Army and the Saudi Arabian National Guard, among others.

| | |
|---|---|
| Country of origin: | Belgium |
| Calibre: | 5.56 x 45mm NATO |
| Length: | 694mm (27.32in) |
| Weight: | 3.6kg (7.93lb) empty, standard configuration |
| Barrel: | 400mm (15.75in) |
| Feed/magazine capacity: | 30-round detachable box magazine |
| Operation: | Gas |
| Cyclic rate of fire: | 850rpm |
| Muzzle velocity: | 900mps (2953fps) |
| Effective range: | 500m (1640ft) |

# QBZ-95

Having lagged behind the rest of the world in terms of assault rifle design, in the 1990s China revealed the QBZ-95 as a new generation of infantry firepower. The impetus behind the rifle was the development of a 5.8 x 42mm cartridge during the late 1980s, which Chinese designers claimed had superior performance to its Western rival, the 5.56 x 45mm NATO. The QBZ-95 was one of a family of weapons created to take the new cartridge. It is of bullpup layout and is a gas-operated, rotating-bolt rifle. The carrying handle at the top of the gun also incorporates an integral rear sight, although the gun can also take external optical or night-vision sights. Underbelly fitment includes a bayonet or a grenade launcher. Other members of the family include sniper, carbine and light support (bipod-mounted) weapons.

| | |
|---|---|
| Country of origin: | China |
| Calibre: | 5.8 x 42mm |
| Length: | 760mm (29.92in) |
| Weight: | 3.4kg (749lb) |
| Barrel: | 520mm (20.47in) |
| Feed/magazine capacity: | 30-round detachable box magazine |
| Operation: | gas, rotating bolt |
| Cyclic rate of fire: | 650rpm |
| Muzzle velocity: | n/a |
| Effective range: | 500m (1640ft) |

# QBZ-03

Disappointment with the QBZ-95 caused Chinese designers to return to a more conventional weapon derived from the Type 81 in order to produce the QBZ-03. Made of a forged aluminium alloy and a polymer compound, the rifle has a gas block with a two-position regulator, one for firing standard ammunition, the other to allow it to fire rifle grenades without the need of an adaptor. Designed to be easy to use for soldiers familiar with previously issued machine guns and rifles, the QBZ-03 has a hooded front sight with a flip up rear diopter similar to the American M16 rifle. The cyclic rate on the issued model is semi-automatic or fully automatic only, while the export model, which, unlike the domestic model is chambered in 5.56x45mm NATO, has an integrated three round burst mode.

| | |
|---|---|
| **Country of origin:** | China |
| **Calibre:** | 5.8mm (.228in) DBP87, 5.56mm (.219in) NATO |
| **Length:** | 960mm (37.79in) stock extended; 710mm (27.95in) stock folded |
| **Weight:** | 3.5kg (7.71lb) |
| **Barrel:** | Not known |
| **Feed/magazine capacity:** | 30-round detachable box |
| **Operation:** | Gas-operated, rotating bolt |
| **Cyclic rate of fire:** | 650rpm |
| **Muzzle velocity:** | 930m/sec (3050ft/sec) |
| **Effective range:** | 400m (1312ft) |

# Samonabiject Puska vz52

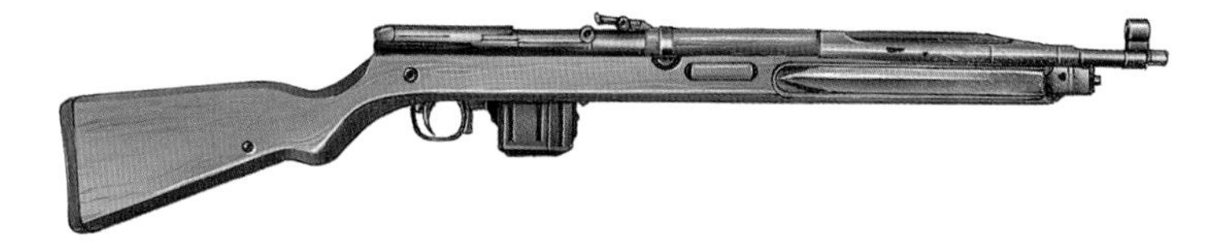

Produced in the immediate aftermath of World War II, the vz52 was a self-loading rifle originally chambered for a new cartridge – the 7.62 x 45mm M52 – which was inspired by the German Kurz cartridge. The vz52 used a gas-operation system in which the gas piston was actually wrapped around the barrel. It also featured a tipping bolt system of locking and was visually distinguished by the attached bayonet which folded into a recess in the forestock. In combination, all these features worked perfectly well and, with the trigger system copied from the US M1 Garand rifle, the vz52 was a perfectly serviceable weapon, if a little heavy. The quality was rather spoiled, however, after the Soviet takeover of Czechoslovakia – the gun was rechambered for the Soviet 7.62 x 39mm M1943 and subsequently became a less dependable weapon.

| | |
|---|---|
| **Country of origin:** | Czechoslovakia |
| **Calibre:** | 7.62 x 45mm M52 or 7.62 x 39mm Soviet M1943 |
| **Length:** | 843mm (33.2in) |
| **Weight:** | 3.11kg (6.86lb) |
| **Barrel:** | 400mm (15.8in), 4 grooves, rh |
| **Feed/magazine capacity:** | 30-round detachable box magazine |
| **Operation:** | Gas |
| **Cyclic rate of fire:** | Semi-automatic |
| **Muzzle velocity:** | 710mps (2330fps) |
| **Effective range:** | 400m (1312ft) |

# Valmet M76

The Valmet M76 is of Soviet AK-47 derivation, but takes advantage of new materials and manufacturing processes, as well as improving on some of the AK-47's strengths. The furniture is the most updated feature of the weapon. The butt comes in a variety of fixed or folding configurations, most particularly a tubular steel variety (the M76T); it also comes in plastic or wooden materials to save weight. At the front of the gun is a plastic forestock and the trigger unit is detachable so that the gun can be fired by a user wearing bulky Arctic mittens, an essential feature for the local climate. A squad automatic version of the M76, the M78, has also been produced, which features a bipod and heavy barrel for light-support fire roles. Being closely derived from the AK weapons, the M76 is a predictably rugged and dependable gun and is capable of hard use.

| | |
|---|---|
| Country of origin: | Finland |
| Calibre: | 7.62 x 39mm Soviet M43; 5.56 x 45mm |
| Length: | 914mm (35.98in) |
| Weight: | 3.6kg (7.94lb) |
| Barrel: | 420mm (16.53in), 4 grooves, rh |
| Feed/magazine capacity: | 15-, 20- or 30-round detachable box magazine |
| Operation: | Gas |
| Cyclic rate of fire: | 650rpm |
| Muzzle velocity: | 720mps (2362fps) |
| Effective range: | 500m (1640ft) plus |

# Fusil MAS Mle 1936

**During the interwar period, French munitions experts sought to develop a cartridge to replace the outdated rimmed 8mm Lebel round. The result was the rimless 7.5 x 54mm Mle 1929. The MAS Mle 1936 was one of the weapons designed to take this new cartridge and has the questionable honour of being the last bolt-action weapon to become a standard military service rifle. Despite a rather awkward appearance, the Mas Mle 1936 was a robust and sound weapon. The bolt action was especially short on account of locking taking place at the very end of the receiver, even though this did necessitate the bolt handle being angled sharply forwards for proximity to the operator's trigger hand. The rifle's production life numbered nearly 20 years (some were adapted for use with rifle grenades in the 1950s) and it also came in a folding butt version for airborne troops.**

| | |
|---|---|
| **Country of origin:** | France |
| **Calibre:** | 7.5mm Mle 1929 |
| **Length:** | 1020mm (40.16in) |
| **Weight:** | 3.78kg (8.33lb) |
| **Barrel:** | 573mm (22.56in), 4 grooves, lh |
| **Feed/magazine capacity:** | 5-round integral box magazine |
| **Operation:** | Bolt action |
| **Muzzle velocity:** | 823mps (2700fps) |
| **Effective range:** | 1000m (3250ft) plus |

# MAS 49/56

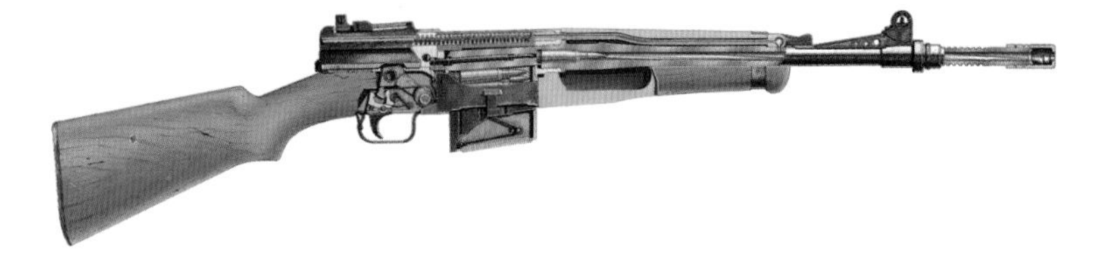

The MAS 49 was France's attempt, in 1949, to give its infantry a self-loading rifle. Apart from the drawback of its unorthodox 7.5mm French Service round (only a small number were produced in 7.62mm NATO), the rifle was distinguished by its almost unfailing reliability due to solid manufacture (even if this did make it a little heavy for the infantryman to carry). It was externally reminiscent of the earlier MAS 39 rifle, but it worked through a gas-operation system in which the bolt was driven back directly by the gas blast, rather than through the use of a piston. The MAS 49 was followed into service by the MAS 49/56. This differed from the original weapon in that, whereas the original's muzzle was designed to accept hollow-tailed rifle grenades, the 49/56 had a flared grenade-launcher/muzzle brake and more exposed barrel.

| | |
|---|---|
| Country of origin: | France |
| Calibre: | 7.5mm French Service |
| Length: | 1010mm (39.76in) |
| Weight: | 3.9kg (8.60lb) |
| Barrel: | 521mm (20.51in), 4 grooves, rh |
| Feed/magazine capacity: | 10-round detachable box magazine |
| Operation: | Gas |
| Cyclic rate of fire: | Semi-automatic |
| Muzzle velocity: | 817mps (2680fps) |
| Effective range: | 500m (1312ft) |

# FAMAS

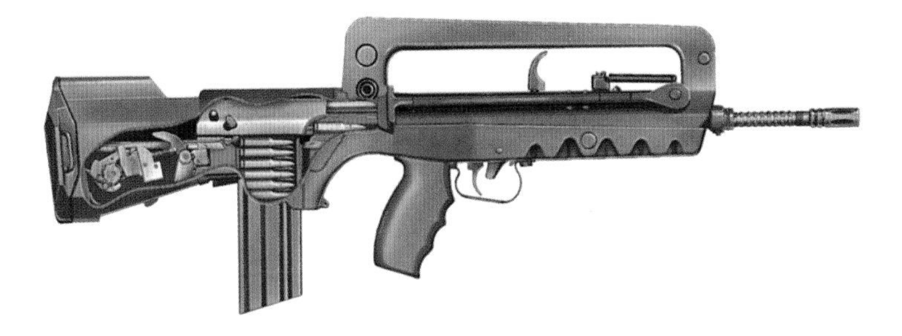

The FAMAS is now the French forces' standard side arm and it is an all-round excellent weapon. Firing standard NATO and French Service 5.56mm ammunition, its bullpup design (with the chamber behind the trigger) gives it good accuracy over 400m (1312ft) due to the length of its barrel, while its light delayed-blowback operation which uses a two-part bolt permits a very high rate of fire – 900rpm can empty the magazine in seconds – something which the trigger's lightness does nothing to control. The FAMAS started to enter French service in the early 1980s and it has given French troops a world-class firearm. The latest model, the F2, has dispensed with the trigger guard in favour of a full handguard ideal for use with gloved hands and it will also take the M16 magazine, now almost an essential feature of NATO weapons.

| | |
|---|---|
| Country of origin: | France |
| Calibre: | 5.56mm NATO or Type France |
| Length: | 757mm (29.8in) |
| Weight: | 3.61kg (7.96lb) |
| Barrel: | 488mm (19.21in), 3 grooves, rh |
| Feed/magazine capacity: | 25-round detachable box magazine |
| Operation: | Gas |
| Cyclic rate of fire: | 900–1000rpm |
| Muzzle velocity: | 960mps (3150fps) |
| Effective range: | 400m (1312ft) |

# Fallschirmjägergewehr 42

**The 7.92mm Fallschirmjägergewehr 42 (FG 42) was developed specifically for the German parachute regiments at the request of the Luftwaffe, which was competing with the Wehrmacht in the early 1940s to produce an automatic rifle for its troops. Rheinmetall subsequently produced the FG 42, which was an innovative and professional weapon. Distinctive features included a gas-operated mechanism, a 20-round side-mounted magazine and a large plastic stock (later, as the war progressed, made of wood because of production problems). Capable of both semi- and full-automatic fire (at 750rpm), it could be used as an assault rifle and as a light infantry support machine gun, the latter function made possible by a permanently fitted bipod near the muzzle. Its in-line stock made it a precursor of many modern assault rifles.**

| | |
|---|---|
| Country of origin: | Germany |
| Calibre: | 7.92mm Mauser |
| Length: | 940mm (37in) |
| Weight: | 4.53kg (9.99lb) |
| Barrel: | 502mm (19.76in), 4 grooves, rh |
| Feed/magazine capacity: | 20-round detachable box magazine |
| Operation: | Gas |
| Cyclic rate of fire: | 750rpm |
| Muzzle velocity: | 761mps (2500fps) |
| Effective range: | 400m (1312ft) plus |

# Machinenpistole 43/Sturmgewehr 44

**The MP43 in every sense marked the genesis of the modern assault rifle. Gas operated, it fired the short and fairly low-powered 7.92 x 33mm Kurz cartridge which was deemed perfectly adequate for the 400m (1312ft) ranges typical of actual combat and enabled the gun to be a stable platform for full-automatic fire. Ironically, Hitler originally prohibited the gun's development when it was known as the Maschinenkarabiner 42 (H), so its name was changed to the MP43 to mask its continuance from Hitler. Once it had proved itself on the Russian Front and Hitler finally gave his blessing, it became known as the Sturmgewehr 44. The impressive, controllable firepower afforded by the MP43/StG.44 impressed all sides during the war and, immediately after 1945, it became the inspiration for many of the modern assault rifles soldiers hold today.**

| | |
|---|---|
| Country of origin: | Germany |
| Calibre: | 7.92mm Kurz |
| Length: | 940mm (37in) |
| Weight: | 5.1kg (11.24lb) |
| Barrel: | 418mm (16.5in), 4 grooves, rh |
| Feed/magazine capacity: | 30-round detachable box magazine |
| Operation: | Gas |
| Cyclic rate of fire: | 500rpm |
| Muzzle velocity: | 700mps (2300fps) |
| Effective range: | c.300m (984ft) |

# Heckler & Koch PSG 1

The Heckler & Koch G3 rifle has already spawned two sniper weapons – the Heckler & Koch G3 A3ZF and the G3 SG/1 – which are only marginally different from the standard gun. The PSG 1 (Präzisionsschützengewehr – High Precision Rifle) features the G3's roller-locked delayed-blowback system; however, it is actually a complete reworking of the G3 design for precision sniping. The barrel itself is a meticulous heavy-duty and extra-long affair featuring a polygonal-rifled bore. This, when used in tandem with the PSG 1's 6x42 telescopic sight which features illuminated cross hairs, gives superb accuracy up to 600m (1968ft), the sight's maximum range configuration with the lowest of its six settings being 100m (328ft). The PSG 1 stock and cheek rest are adjustable in both length and height, and the model can also be fitted with a lightweight bipod.

| | |
|---|---|
| **Country of origin:** | Germany |
| **Calibre:** | 7.62 x 51mm NATO |
| **Length:** | 1208mm (47.56in) |
| **Weight:** | 8.1kg (17.86lb) |
| **Barrel:** | 650mm (25.6in), 4 grooves, rh |
| **Feed/magazine capacity:** | 5- or 20-round detachable box magazine |
| **Operation:** | Roller-locked delayed blowback |
| **Cyclic rate of fire:** | Semi-automatic |
| **Muzzle velocity:** | 815mps (2675fps) |
| **Effective range:** | 600m (1968ft) |

# Heckler & Koch G3

Although the G3 does not have the fame of the M16 or the AK47, it still stands as one of the most widely distributed modern assault rifles, used by more than 50 armies. Based on a CETME design, its operation is a roller delayed blowback with a heritage that can be traced to Mauser in the mid-1940s. Its rollers move in and out of recesses in the receiver wall and this rugged operation makes the G3 a dependable rifle, firing the powerful 7.62mm NATO cartridge. It is arguable whether using the full-size rifle cartridge has done the G3 any favours; on the one hand, it gave the rifle great range and power, but on the other it made the rifle date quickly once the new 5.56mm weapons appeared. The gun could be rather heavy and had a crude appearance due to extensive use of metal stampings. The G3 has been thoroughly combat-tested in theatres ranging from Africa and the Middle East to South America.

| | |
|---|---|
| Country of origin: | Germany |
| Calibre: | 7.62mm NATO |
| Length: | 1025mm (40.35in) |
| Weight: | 4.4kg (9.7lb) |
| Barrel: | 450mm (17.71in), 4 grooves, rh |
| Feed/magazine capacity: | 20-round detachable box magazine |
| Operation: | Delayed blowback |
| Cyclic rate of fire: | 500–600rpm |
| Muzzle velocity: | 800mps (2625fps) |
| Effective range: | 500m (1650ft) plus |

# Heckler & Koch G3SG1

The G3 has many variants, including a version in 5.56mm calibre, the HK 33 – which at one stage looked like being adopted by the German Army as their standard rifle – and folding stock models. It also came in its own sniper version known as the Scharfschützen Gewehr, or SG-1. This was essentially a standard G3 rifle but featured a telescopic sight (usually a Schmidt & Bender) and also a more sensitive trigger unit and specially selected barrel. It has a lightweight bipod attached to the front of the foregrip. In this form the G3 entered service as a police weapon. The rifle itself is identical to the G3, using a minimum of expensive machined parts, and like the Lee Enfield Enforcers in British police service, continue to give good results due to the rugged nature of the design, which can withstand rough treatment.

| | |
|---|---|
| Country of origin: | Germany |
| Calibre: | 7.62mm NATO |
| Length: | 1025mm (40.35in) |
| Weight: | 4.4kg (9.7lb) |
| Barrel: | 450mm (17.71in), 4 grooves, rh |
| Feed/magazine capacity: | 20-round detachable box magazine |
| Operation: | Delayed blowback |
| Cyclic rate of fire: | 500–600rpm |
| Muzzle velocity: | 800mps (2625fps) |
| Effective range: | 500m (1650ft) plus |

# Heckler & Koch G11

The Hecker & Koch G11 gives us a glimpse of the future of individual weaponry. Scarcely looking like a gun at all, it fires caseless rounds which consist of rectangular blocks of propellant with the bullet and percussion cap set in. The advantages of this system include reduced cost and weight, no ejection and extraction stoppages, more economical production and smaller cartridge dimensions. The rounds are fed from a horizontal magazine attached above the barrel. In full automatic mode, 600rpm is achieved; in three-round burst, this figure climbs to 2200rpm, the rate increase being achieved by all three rounds being fired in one recoil cycle. The three-round speed enables great accuracy with climb almost eliminated. Although the G11 was intended as the German Army's standard rifle replacement in 1990, internal disputes have kept it the preserve of special forces.

| | |
|---|---|
| Country of origin: | Germany |
| Calibre: | 4.7 x 33mm DM11 caseless |
| Length: | 752.5mm (29.62in) |
| Weight: | 3.80kg (8.38lb) |
| Barrel: | 537.5mm (21.16in), 6 grooves, polygonal, rh |
| Feed/magazine capacity: | 50-round detachable box magazine |
| Operation: | Gas |
| Cyclic rate of fire: | 600rpm (full automatic); 2200rpm (three-round burst) |
| Muzzle velocity: | 930mps (3050fps) |
| Effective range: | 500m (1640ft) plus |

# Heckler & Koch G36

**The G36 is a recent H&K development intended to take over from the HK33 assault rifle. It was unveiled in 1997, though details are still sketchy about production plans. In it H&K have moved away from the roller-locked delayed-blowback operation of their previous assault rifles to a gas-operated system of a type similar to that used in the Armalite AR-18. The G36 comes in three versions: a standard G36 rifle with bipod, the G36K carbine with more compact dimensions for vehicular/special forces use, and the MG36 which is actually a heavy-barrelled light support weapon. Several European countries including Germany, Spain and Norway are considering adopting the G36 as a standard weapon, and by most accounts it is an excellent, accurate weapon. In common with most Heckler & Koch products, it is likely to see widespread service.**

| | |
|---|---|
| Country of Origin: | Germany |
| Calibre: | 5.56 x 45mm NATO |
| Length: | 999mm (39.3in) stock extended; 758mm (29.8in) stock folded |
| Weight: | 3.4kg (7.49lb) |
| Barrel: | 480mm (18.9in), 6 grooves, rh |
| Feed/Magazine capacity: | 30-round detachable box magazine |
| Operation: | Gas operated |
| Cyclic rate of fire: | 750rpm |
| Muzzle Velocity: | Not available |
| Effective Range: | Not available |

# Heckler & Koch HK416

**Based on the US M4 Carbine, the HK416 uses a gas piston system derived from the G36 rifle. It has four picatinny rails for a range of accessories. The rail forearm can be installed and removed without tools by using the bolt locking lug as the screwdriver. The gun has an adjustable multi-position telescopic butt stock, offering six different lengths of pull. The shoulder pad can be either convex or concave and the stock features a storage space for maintenance accessories, spare electrical batteries or other small kit items. The gun requires a thicker barrel than the M4 because its short stroke gas piston system produces more pressure. In 2004, the US Delta Force replaced its M4s with HK416s after tests showed that the Heckler & Koch malfunctioned less and caused less wear on parts. The weapon is also used by the Norwegian armed forces.**

| | |
|---|---|
| Country of origin: | Germany |
| Calibre: | 5.56mm (0.21in) |
| Length: | 690mm (27.2in) |
| Weight: | 2.950kg (6.50lb) |
| Barrel: | 228mm (9.0in) |
| Feed/magazine capacity: | 20-, 30-round STANAG magazine or 100-round Beta C-Mag box magazine |
| Operation: | Gas, rotating bolt |
| Cyclic rate of fire: | 700–900 rounds/min (cyclic) HK416 |
| Muzzle velocity: | Varies by barrel length and type of round used |
| Effective range: | 365m (1200ft) |

# Walther WA2000

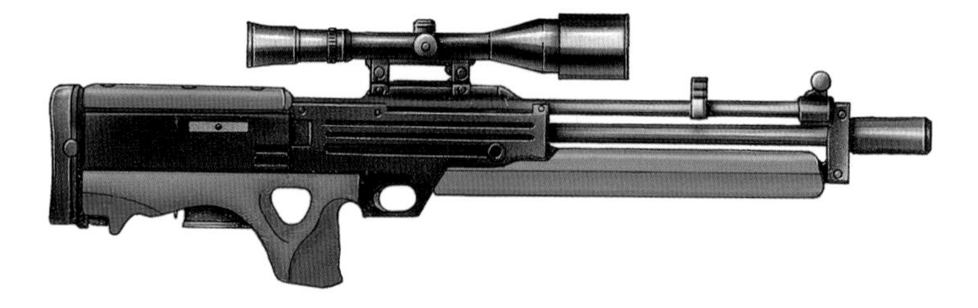

The Walther WA2000 represents a revolution in sniper weapon design. The barrel is clamped at the front and rear so that it does not twist on firing, while the rest of the barrel is free from any contact with furniture to prevent further distortions or movement disturbing the gun's aim. Furthermore, the barrel is fluted to dampen vibrations and aid cooling. The barrel is set in line with the user's shoulder to reduce recoil and the bolt mechanism sits behind the handgrip in a 'bullpup' arrangement. All stock furniture is fully adjustable and, when firing the .300 Winchester Magnum and using the standard Schmidt and Bender x2.5 to x10 sight, the rifle's accuracy is considerable. Its sophistication and price tag, however, mean that it is more suited to police and security operations than arduous military use in the field.

| | |
|---|---|
| Country of origin: | Germany |
| Calibre: | .300in (7.62m) Winchester Magnum |
| Length: | 905mm (35.63in) |
| Weight: | 8.31kg (18.32lb) loaded, with telescopic sight |
| Barrel: | 650mm (25.59in) |
| Feed/magazine capacity: | 6-round detachable box magazine |
| Operation: | Gas |
| Muzzle velocity: | c.800mps (2624fps) |
| Effective range: | 1000m (3250ft) plus |

# L1A1 Self-loading Rifle

The persuasive qualities of the Belgian FN FAL rifle led the British Army to adopt it as their standard service rifle from 1954, but licence-produced as the L1A1 Self-loading Rifle. It was little different from the FN FAL original, although the L1A1 had its dimensions slightly altered to suit manufacture in imperial measurements and the British rifle would only fire semi-automatic. The L1A1 served British soldiers well across the world. On the streets of Northern Ireland, it was over-powerful, as it could easily command ranges of more than 800m (2600ft), far in excess of those combat ranges required in the primarily urban setting. Yet it was perhaps the Falklands War when the L1A1 left its mark, when it could easily handle the distances between Argentine and British positions, and also had good stopping power. It was used by special forces units in the Gulf War for its long range.

| | |
|---|---|
| Country of origin: | Great Britain |
| Calibre: | 7.62 x 51mm NATO |
| Length: | 1055mm (41.5in) |
| Weight: | 4.31kg (9.5lb) |
| Barrel: | 535mm (21in), 4 grooves, rh |
| Feed/magazine capacity: | 20-round box magazine |
| Operation: | Gas |
| Cyclic rate of fire: | Semi-automatic only |
| Muzzle velocity: | 853mps (2800fps) |
| Effective range: | 800m (2600ft) plus |

# EM-2 Assault Rifle

The story of the EM-2 Assault Rifle is that of a gun truly ahead of its time. It was designed in the United Kingdom shortly after World War II, as a replacement for the over-powerful .303 Lee-Enfield rifle, and intended to be the new standard British infantryman's weapon. The EM-2 was revolutionary in that it fired a 7mm short-cased cartridge with magazine and bolt system behind the trigger group in what is now known as a 'bullpup' layout. The result was surprisingly good. It was an accurate (it fired from a closed bolt), durable weapon which became destined for British Army service as Rifle, Automatic, 7mm No. 9 Mk. 1.; however, the USA objected to both the cartridge size and power, and the 7.62mm round was accepted as the NATO standard. The EM-2 thus had to be dropped, although history has shown that the principles of the EM-2 would be followed by even its critics.

| | |
|---|---|
| Country of origin: | Great Britain |
| Calibre: | 7mm British |
| Length: | 889mm (35in) |
| Weight: | 3.41kg (7.52lb) |
| Barrel: | 623mm (24.5in), 4 grooves, rh |
| Feed/magazine capacity: | 20-round detachable box magazine |
| Operation: | Gas |
| Cyclic rate of fire: | 600–650rpm |
| Muzzle velocity: | 771mps (2530fps) |
| Effective range: | 400m (1312ft) plus |

# Enfield Individual Weapon L85A1 (SA80)

**The L85A1 (otherwise known as the SA80) is one of those firearms which is sound in theory and practice, but which has been badly let down in active-service conditions. Built by the Royal Small Arms Factory, it replaced the L1A1 Self-Loading Rifle (the British variant of the FN FAL rifle) as the British Army's standard infantry weapon in the early 1980s. Gas-operated and working on a 'bullpup' design which places the magazine behind the trigger (thus maximising barrel length), the L85A1 is accurate and has very little recoil. First-line troops have the weapon fitted with SUSAT (Sight Unit, Small Arm, Trilux) optical sight, which further enhances the accuracy of use. Its undoubtedly fine design qualities are let down, however, by both engineering defects in the early models and a lack of durability under even moderately dirty conditions.**

| | |
|---|---|
| **Country of origin:** | Great Britain |
| **Calibre:** | 5.56mm NATO |
| **Length:** | 785mm (30.9in) |
| **Weight:** | 3.80kg (8.3lb) |
| **Barrel:** | 518mm (20.39in), 6 grooves, rh |
| **Feed/magazine capacity:** | 30-round detachable box magazine |
| **Operation:** | Gas |
| **Cyclic rate of fire:** | 700rpm |
| **Muzzle velocity:** | 940mps (3084fps) |
| **Effective range:** | 400m (1312ft) |

# Accuracy International AS50

Since the 1990s, .50 BMG (12.7 x 99mm) sniper rifles have become increasingly popular as long-range anti-materiel/personnel weapons. The AS50 is a semi-auto rifle in this calibre produced in the UK by the Accuracy International company. It is gas-operated, the massive recoil being controlled by the sheer size of the gun, heavy buffer springs and a sizeable muzzle brake. (The bolt-action version of this rifle, the AW50, has a higher recoil impact.) The magazine holds five rounds and the rifle comes with a folding bipod and rear support leg as standard. Disassembly has been made a simple affair, taking less than three minutes and not requiring special tools. Allied to an appropriate sight, the AS50 can deliver accurate shots to and exceeding 1500m (4921ft), and those struck by the bullet are unlikely to survive the experience.

| | |
|---|---|
| **Country of origin:** | Great Britain |
| **Calibre:** | 12.7 x 99mm (.50 BMG) |
| **Length:** | 1369mm (53.89in) |
| **Weight:** | 14.1kg (31lb) |
| **Barrel:** | 692mm (27.24in) |
| **Feed/magazine capacity:** | 5-round box magazine |
| **Operation:** | Gas |
| **Cyclic rate of fire:** | n/a |
| **Muzzle velocity:** | 843mps (2800fps) |
| **Effective range:** | 1500m (4921ft) plus |

# INSAS Assault Rifle

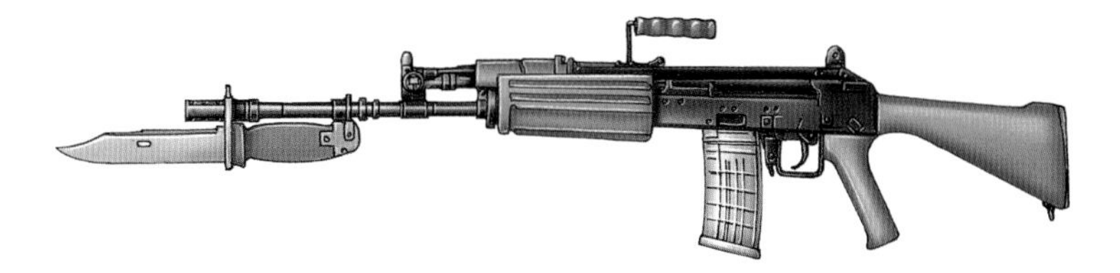

This indigenous Indian weapon from the Indian Small Arms Factory at Kanpur has given Indian troops a thoroughly modern assault rifle through a process of adapting existing designs to Indian production. The three most seminal assault rifle series of the 20th century – the AK, the M16 and the Heckler &Koch – are present in various elements of the gun and combined to make a reliable, capable gas-operated rifle. Unlike many of these other weapon series, the INSAS is not capable of fully automatic fire – only single shots or three-round burst options are available. This shows a maturity and combat-mindedness on the part of the designers, as fully automatic fire can simply allow the user to empty his rifle inaccurately during a firefight, whereas more limiting types of fire seem to encourage accurate aim by the weapon's users.

| | |
|---|---|
| Country of origin: | India |
| Calibre: | 5.56 x 45mm NATO |
| Length: | 990mm (38.97in) |
| Weight: | 3.2kg (7.05lb) |
| Barrel: | 464mm (18.26in) |
| Feed/magazine capacity: | 20- or 30-round detachable box magazine |
| Operation: | Gas, self-loading |
| Cyclic rate of fire: | Semi-automatic |
| Muzzle velocity: | 985mps (2903fps) |
| Effective range: | 800m (2600ft) |

# Pindad SS2

The Pindad SS2 is derived from the FN FNC rifle. It is actually a weapon family, with carbine, rifle and 'para-sniper' versions all based on a common receiver. The carrying handle and detachable rear sight is placed on top of Picatinny railings. The carrying handle is replaced with a Picatinny rail for scope mounting as a designated marksman rifle on the SS2-V4 variant. The flash suppressor and forward assist is based on the Colt M16A2 with the front sight being based on the AK rifles. The weapon shown has been fitted with an SPG-1A underbarrel grenade launcher. They are in service with both the Indonesian police, the Indonesian Army and Komando Pasukan Katak, the frogman and underwater demolition unit of the Indonesian Navy, and Indonesian police.

| | |
|---|---|
| Country of origin: | Indonesia |
| Calibre: | 5.56mm (0.219in) NATO |
| Length: | 990mm (38.97in) |
| Weight: | 3.4kg (7.49lb) |
| Barrel: | 740mm (29.13in) |
| Feed/magazine capacity: | Various STANAG magazines |
| Operation: | Gas |
| Cyclic rate of fire: | 650–675rpm |
| Muzzle velocity: | 710m/sec (2329ft/sec) |
| Effective range: | 500m (1640ft) |

# Khaybar KH 2002

Developed from the M16 by way of the Chinese CQ 5.56 rifle, the Iranian Khaybar may have found its way into the hands of fighters resisting the US-led Coalition occupation of Iraq in 2004, despite efforts to prevent arms smuggling in the region. The gun can be selected to fire three-round bursts or be used as an automatic or semi-automatic. The fire selector lever can be seen on the left side behind the magazine. Night sights or opticals can be mounted on the carrying handle, which also carries the rear sight. A bayonet or bipod, as illustrated, can also be fitted. The bolt mechanism could be flipped to allow for left-handed firing, making the KH 2002 ambidextrous. Developed from the basic assault rifle are also a shortened-barrel assault carbine form and a longer-barrelled marksman form with specialized optics.

| | |
|---|---|
| Country of origin: | Iran |
| Calibre: | 5.56mm (.219in) |
| Length: | 730mm (28.7in) |
| Weight: | 3.7kg (8.15lb) |
| Barrel: | Not known |
| Feed/magazine capacity: | Various STANAG magazines |
| Operation: | Gas, rotating bolt |
| Cyclic rate of fire: | 800–850rpm, cyclic |
| Muzzle velocity: | 900–950mps (2952–3116fps) |
| Effective range: | 450m (1476ft) |

# Galil ARM

The Galil was designed as a lighter replacement for the FN FAL rifle for the Israeli Defence Force (IDF) in the wake of the 1967 Six-Day War. After an intense period of study of the world's assault rifles, Israeli Military Industries (IMI) opted for a version of the AK-series rotating-bolt action and the Galil was initially built using the body of another AK derivative, the Finnish M62 Valmet. The standard rifle is the Galil ARM, which is fitted with a bipod and carrying handle, and can be used in a light machine-gun role; the Galil AR lacks both bipod and carrying handle; and the Galil SAR (Short Assault Rifle) is the same as the AR, but with a shorter barrel. An excellent gun, a version of the Galil was made for 7.62mm calibre, but the 5.56mm version is dominant in Israel and other international armies. All Galils come with either folding or fixed butts.

| | |
|---|---|
| Country of origin: | Israel |
| Calibre: | 5.56mm NATO |
| Length: | 979mm (38.54in) overall; 742mm (29.21in) stock folded |
| Weight: | 4.35kg (9.59lb) |
| Barrel: | 460mm (18.11in), 6 grooves, rh |
| Feed/magazine capacity: | 35- or 50-round box magazine |
| Operation: | Gas, self-loading |
| Cyclic rate of fire: | 650rpm |
| Muzzle velocity: | 990mps (3250fps) |
| Effective range: | 800m (2624ft) plus |

# Galil Sniper

The Galil Sniper rifle emerged as a redesign by Israel Military Industries (IMI) of the standard Galil service rifle, but that redesign has been exceptionally thorough. In addition to a heavy barrel and bipod, the Galil Sniper differs from the service gun by having an adjustable stock (with shoulder and cheek adjustments) which folds down for storage, a muzzle brake to limit recoil and a Nimrod x6 telescopic sight (fitted to the left side of the receiver). In addition, almost all the internal workings have been exhaustively retuned for the gun's accurate role and the weapon is only capable of semi-automatic fire. Having been born from the service rifle has also ensured that the gun is robust enough to handle military life. In trials and in combat, the Galil sniper has shown its lethal accuracy up to, and beyond, ranges of 800m (2624ft).

| | |
|---|---|
| Country of origin: | Israel |
| Calibre: | 7.62mm NATO |
| Length: | 1115mm (43.89in) stock extended; 840mm (33in) stock folded |
| Weight: | 6.4kg (14.11lb) |
| Barrel: | 508mm (20in) without muzzle brake, 4 grooves, rh |
| Feed/magazine capacity: | 20-round box magazine |
| Operation: | Gas, self-loading |
| Muzzle velocity: | 815mps (2675fps) |
| Effective range: | 800m (2624ft) plus |

# IMI Tavor TAR-21

Although the Israel Defense Forces (IDF) have been well-served by rifles such as the M16 and Galil, a replacement rifle was developed in the 1990s by Israel Military Industries (IMI) and named the Tavor TAR-21. The Tavor has since been adopted into Israeli service, although many legacy weapons remain. In basic structure, the TAR-21 is a gas-operated bullpup rifle fitted – like many similar weapons – with an optical sight as standard. It takes the principle of modularity to new levels, however. It can be fitted with the M203 grenade launcher and grenade-launching sight; used with a bipod to create the STAR 21 (designated marksman) rifle; it is also produced in a more compact CTAR assault version. There is even a micro version, the MTAR, which can be converted to 9mm ammunition.

| | |
|---|---|
| Country of origin: | Israel |
| Calibre: | 5.56 x 45mm NATO |
| Length: | 720mm (28.35in) |
| Weight: | 3.27kg (7.21lb) |
| Barrel: | 460mm (18.11in) |
| Feed/magazine capacity: | 30-round detachable box magazine |
| Operation: | gas |
| Cyclic rate of fire: | 750–900rpm |
| Muzzle velocity: | 910mps (2986fps) |
| Effective range: | 550m (1804ft) |

# Beretta BM59

The BM59 grew out of the US M1 Garand rifle, which Beretta made under licence in Italy following World War II. The redesign took place in 1959, when NATO members were standardising weapons to the 7.62mm calibre. The most fundamental differences between the M1 and BM59 were that the BM59 now had an extended 20-round magazine and it could be selected to fire full-auto. The changes to achieve this configuration were actually minimal, and the M1's gas-operation system remained essentially the same. Once the BM59 was established, other models were brought out with varied butt, barrel, bayonet and grenade-launching adaptations for use by certain specialist units such as airborne and mountain troops. A heavy barrelled version was also made for use in a sustained-fire role. Like the other M1 redevelopment, the M14, the BM59 is now somewhat dated in weight and configuration.

| | |
|---|---|
| Country of origin: | Italy |
| Calibre: | 7.62mm NATO |
| Length: | 1095mm (43.11in) |
| Weight: | 4.6kg (10.14lb) |
| Barrel: | 490mm (19.29in), 4 grooves, rh |
| Feed/magazine capacity: | 20-round box |
| Operation: | Gas |
| Cyclic rate of fire: | 750rpm |
| Muzzle velocity: | 823mps (2700fps) |
| Effective range: | 800m (2600ft) |

# Beretta AR70/90

The AR70/90 was developed during the 1980s and brought into production in 1990 as a much improved version of its predecessor, the AR70/223. The fault of the AR70/223 lies in the weak design of its receiver, which could precipitate bolt jamming, so in the AR70/90 the receiver has been redesigned, with more reliable bolt guides in place. The AR70/90 can fire in semi- and full-automatic modes and three-round bursts, and its variant models alter such features as barrel length and stock configuration (folding or fixed) according to usage. The SCS-70/90 model has a folding butt and 320mm (12.5in) barrel. It can accept standard M16 magazines and it has a carrying handle/sight situated above the receiver, in contrast to the AR70/223. The AR70/90 is the standard Italian Army rifle and has also been exported to Egypt, Morocco, Jordan and Zimbabwe.

| | |
|---|---|
| Country of origin: | Italy |
| Calibre: | 5.56mm NATO |
| Length: | 998mm (39.29in) |
| Weight: | 3.99kg (8.8lb) |
| Barrel: | 450mm (17.72in), 6 grooves, rh |
| Feed/magazine capacity: | 20- or 30-round box magazine |
| Operation: | Gas, self-loading |
| Cyclic rate of fire: | 650rpm |
| Muzzle velocity: | 950mps (3116fps) |
| Effective range: | 500m (1650ft) |

# Beretta AS70/90

**Beretta's excellent AR70/90 assault rifle spawned an extensive range of derivative weapons, including the AS70/90, which is intended to partner the AR70/90 in service. This weapon is a heavy barrelled version of the rifle – it uses the same system of gas-operation as the rifle – and serves as a squad automatic weapon or light machine gun. Its support role is facilitated by an adjustable bipod attached just behind the front sight and for accuracy the carrying handle can be removed to allow fitting of advanced sighting systems. The AS70/90 is also capable of firing rifle grenades from the muzzle. In terms of quality and performance, the status of the AS70/90 cannot be doubted. Yet as it attempts to enter new markets it remains to be seen whether the fact that its barrel does not allow for a quick change in case of overheating will count against it.**

| | |
|---|---|
| Country of origin: | Italy |
| Calibre: | 5.56 x 45mm NATO |
| Length: | 1000mm (39.37in) |
| Weight: | 5.34kg (11.77lb) |
| Barrel: | 465mm (18.3in), 6 grooves, rh |
| Feed/magazine capacity: | 30-round detachable box magazine |
| Operation: | Gas |
| Cyclic rate of fire: | 800rpm |
| Muzzle velocity: | 980mps (3215fps) |
| Effective range: | 500m (1640ft) plus |

# FX-05 Xiuhcoatl

This home-grown Mexican rifle is a recent addition to the list of modern assault rifles. Developed in 2005–06, it was adopted by the Mexican armed forces as an alternative to the Heckler & Koch G36, which was in turn intended to be a replacement for the venerable H&K G3. The Xiuhcoatl – the name means 'Fire Serpent' – is a gas-operated weapon capable of firing either single shot, three-round burst or full auto, the latter at a rate of 750rpm. Weighing 3.9kg (8.59lb), it is nearly 1kg (2.2lb) lighter than the G3, although the fact that it fires the 5.56 x 45mm NATO round rather than the G3's 7.62 x 51mm NATO provides a natural weight advantage. It comes with a Picatinny rail for fitting various scopes and other attachments, has a folding stock, and there are also short-barrelled versions for police and security forces use.

| | |
|---|---|
| Country of origin: | Mexico |
| Calibre: | 5.56 x 45mm NATO |
| Length: | 1087mm (42.79in) |
| Weight: | 3.9kg (8.59lb) |
| Barrel: | n/a |
| Feed/magazine capacity: | 30-round detachable box magazine |
| Operation: | Gas |
| Cyclic rate of fire: | 750rpm |
| Muzzle velocity: | 910mps (2986fps) |
| Effective range: | 550m (1804ft) |

# SAR 80

The SAR 80 was inspired by the US M16 rifle, but, following an intense design liaison between Chartered Industries of Singapore and Sterling Armaments of England, a weapon perhaps even superior to the M16 was born. Like the M16, it is a gas-operated weapon using a straight-in-line design to aid controllability and accuracy, and it can accept the standard 5.56mm M16 magazine. Singapore's investment in new technology, however, has made the SAR 80 a much cheaper weapon to manufacture and orders are already spreading across the world (large numbers were recently seen in Yugoslavia during the civil war.) The SAR 80 can give single, three-round burst or full-automatic fire, and also accepts rifle grenades directly onto its flash-suppressor. These features plus its overall quality mean that the SAR 80 seems destined to be one of Singapore's successful exports.

| | |
|---|---|
| Country of origin: | Singapore |
| Calibre: | 5.56 x 45mm NATO |
| Length: | 970mm (38.18in) |
| Weight: | 3.17kg (7lb) |
| Barrel: | 459mm (18.07in), 6 grooves, rh |
| Feed/magazine capacity: | 30-round detachable box magazine |
| Operation: | Gas, self-loading |
| Cyclic rate of fire: | 700rpm |
| Muzzle velocity: | 970mps (3182fps) |
| Effective range: | 800m (2600ft) plus |

# SR-88

Developed from the SAR-80, the SR-88 owed much of its design to its predecessor: it was a gas-operated weapon with a gas cylinder fitted over the barrel. The firing action relied on a rotating bolt internal system and the weapon was fed from a standard 30-round STANAG-type curved detached box magazine. Like the M16, the weapon was chambered for the 5.56 x 45 NATO standard cartridge. A selector allowed for burst or automatic fire and the gun could take the American M203 under-barrel grenade launcher. The standard SR-88 was produced from 1988 to 1995, when the SR-88A was introduced. This improved version included luminous night sights and an optical mount, along with an optional folding shoulder stock. The SR-88A remained in production until the year 2000. Apart from the Singaporean Armed Forces, the gun was only ever used by the Slovenian Army.

| | |
|---|---|
| Country of origin: | Singapore |
| Calibre: | 5.56mm (.219in) NATO |
| Length: | 960mm (37.7in) |
| Weight: | 3.68kg (8.11lb) |
| Barrel: | 460mm (18.1in) |
| Feed/magazine capacity: | 30-round detachable box |
| Operation: | Gas, rotating bolt |
| Cyclic rate of fire: | 750rpm |
| Muzzle velocity: | Not known |
| Effective range: | 800m (2625ft) |

# Vektor R4

The Vektor R4 is a superb assault rifle based on the Israeli Galil and one which acted as a replacement for the South African Defence Force's FN FALs and Heckler & Koch G3s. The R4 makes greater use of a high-impact nylon/glass fibre mix in its construction than the Galil and is generally stronger and larger than the Israeli gun, although it weighs approximately the same. A bipod fitting comes as standard for the rifle, which also has a wire-cutting feature and a bottle-opener (the latter prevents soldiers opening bottles on precision gun components, such as the magazine receiver lips, and thus damaging them). A further feature is the use of tritium inserts into the sights for night firing. The R4 is at the top of a series of weapons which includes a carbine version, the R5 and the even shorter R6, with its 280mm (1.10in) barrel.

| | |
|---|---|
| Country of origin: | South Africa |
| Calibre: | 5.56 x 45mm M193 |
| Length: | 1005mm (35.97in) stock extended; 740mm (29.13in) stock folded |
| Weight: | 4.3kg (9.48lb) |
| Barrel: | 460mm (18.11in), 6 grooves, rh |
| Feed/magazine capacity: | 35 or 50-round detachable box magazine |
| Operation: | Gas |
| Cyclic rate of fire: | 650rpm |
| Muzzle velocity: | 980mps (3215fps) |
| Effective range: | 500m (1640ft) plus |

# CETME

The CETME takes its rank alongside the Heckler & Koch G3 and the FN FAL as one of the most widely used assault rifles, even though it perhaps did not quite achieve the level of popularity as these competitors. The CETME rifles actually originate from Mauser, which developed a prototype rifle during World War II which turned into a production model when the designers moved to fascist Spain's CETME concern following the end of the war. The CETME Model 58 series was the original. It first fired a 7.92mm round, then a special reduced-power 7.62mm round. Its operation was to be influential: the roller-locked delayed-action blowback came to be used by Heckler & Koch in some of their weapons. The subsequent history of the CETME rifles is one of recalibration. After changing to the full-power NATO round in 1974, the latest models (L and LC) use the standard 5.56mm NATO round.

| | |
|---|---|
| Country of origin: | Spain |
| Calibre: | 7.92 x 51mm; 7.62 x 51mm; 7.62 x 51mm NATO; 5.56 x 45mm NATO |
| Length: | 1015mm (39.96in) |
| Weight: | 4.5kg (9.92lb) |
| Barrel: | 450mm (17.72in), 4 grooves, rh |
| Feed/magazine capacity: | 20-round detachable box magazine |
| Operation: | Gas, self-loading |
| Cyclic rate of fire: | 600rpm |
| Muzzle velocity: | 7.92mm: 780mps (2560fps); 5.56mm: 875mps (2878fps) |
| Effective range: | 800m (2600ft) plus |

# SIG SG540

The SG540 has become a broad export for the Swiss firm SIG, going out to markets across the world (particularly those in Africa, South America and the Middle East) and being manufactured for a short time under licence by Manhurin of France. The gun is a rotating-bolt gas-operated weapon which had its ancestry back in the Stgw 57 assault rifle. It is 5.56mm in calibre along with the SG543 – the SG542 was made for the 7.62 x 51mm NATO round. Its greatest virtues are its reliability and its adaptability through a range of different fittings. A bipod and telescopic sight can make it into an effective sniper rifle, while the flash-compensator doubles as a grenade-launching mount. The SG540 also features a tilted drum rear sight which can give various range settings at 100m (328ft) intervals. Overall the SG540 is an excellent infantry weapon of its type.

| | |
|---|---|
| Country of origin: | Switzerland |
| Calibre: | 5.56 x 45mm NATO |
| Length: | 950mm (37in) |
| Weight: | 3.26kg (7.19lb) |
| Barrel: | 460mm (18in), 6 grooves, rh |
| Feed/magazine capacity: | 20- or 30-round detachable box magazine |
| Operation: | Gas, rotating bolt |
| Cyclic rate of fire: | 650–800rpm |
| Muzzle velocity: | 980mps (3215fps); 5.56mm |
| Effective range: | 800m (2600ft) |

# Browning Automatic Rifle

The Browning Automatic Rifle, or BAR, was something of an oddity, as it was classed as a rifle, but was more of a light machine gun in its dimensions. It was developed by Browning in 1917 and used in World War I in 1918, after which it was fitted with a bipod and became the BAR M1918A1, then the slightly improved BAR M1918A2. The BAR was a gas-operated gun which, depending on model and sub-variations, could fire either full automatic at two rates, 350 or 550rpm, or single-shot and full automatic. Firing full auto from a magazine that only held 20 rounds was somewhat incongruous, but, despite this and its heavy weight, the BAR became immensely popular with US servicemen during World War II and it remained in active service until 1957. In combat during WWII the BAR was often seen with the bipod removed, to save weight.

| | |
|---|---|
| Country of origin: | USA |
| Calibre: | .30in M1906 |
| Length: | 1219mm (48in) |
| Weight: | 8.8kg (19.4lb) |
| Barrel: | 610mm (24in), 4 grooves, rh |
| Feed/magazine capacity: | 20-round detachable box magazine |
| Operation: | Gas |
| Cyclic rate of fire: | 550 or 350rpm |
| Muzzle velocity: | 808mps (2650fps) |
| Effective range: | 800m (2600ft) plus |

# M1 Rifle (Garand)

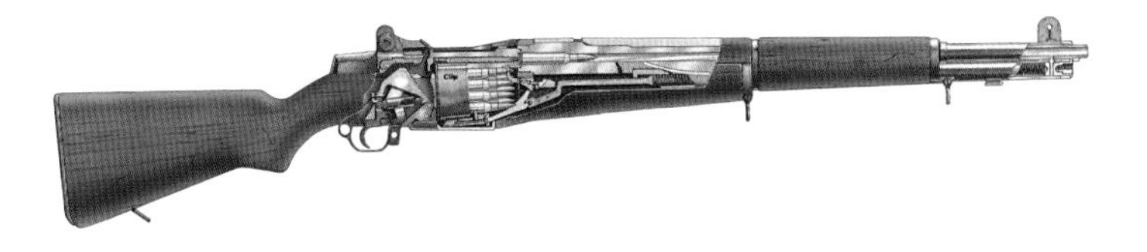

The M1 became almost a symbol of US force during World War II and is the first self-loading rifle to be adopted as a standard military firearm. Designed by John C. Garand, a French-Canadian by birth, and produced from 1936, its main virtues were its self-loading gas operation, its solidity and durability (though it was also heavy), along with its manufacture in enormous numbers – up to six million by the time it ended production in 1959. The negatives of the M1 were that the eight-round clips had to be loaded full or not at all – single rounds could not be added to a clip from which rounds had already been fired – and a loud 'ping' when the magazine emptied its last round which could act as an advertisement of vulnerability to the enemy. (Cunning GIs would throw empty clips to simulate their running out of ammunition, tempting unwary enemy soldiers out of cover).

| | |
|---|---|
| Country of origin: | USA |
| Calibre: | US .30-06 |
| Length: | 1103mm (43.5in) |
| Weight: | 4.37kg (9.5lb) |
| Barrel: | 610mm (24in), 4 grooves, rh |
| Feed/magazine capacity: | 8-round internal box magazine |
| Operation: | Gas |
| Muzzle velocity: | 853mps (2800fps) |
| Effective range: | 500m (1650ft) plus |

# Carbine, Caliber .30, M1

**The M1 Carbine has endured in popularity from its initial production date in 1942 until the present day. It was originally designed by Winchester as a light, portable weapon for second-echelon troops such as drivers and support personnel, but these very qualities led many frontline combatants to adopt the weapon. The M1's distinctive gas-operation system worked through gas pressure pushing back a short-stroke piston, which in turn drove back the operating rod which cycled the bolt. It worked well and the M1 became a popular gun, although its pistol cartridge made it only suitable for relatively close-range fighting. Important variants included the M1A1, which had a folding steel butt, and the M2, which could fire full automatic, and in total more than six million M1 or variants were produced during World War II.**

| | |
|---|---|
| Country of origin: | USA |
| Calibre: | .30in Carbine |
| Length: | 905mm (35.7in) |
| Weight: | 2.5kg (5.47lb) |
| Barrel: | 457mm (18in), 4 grooves, rh |
| Feed/magazine capacity: | 15- or 30-round detachable box magazine |
| Operation: | Gas |
| Cyclic rate of fire: | M2/M3: 750rpm |
| Muzzle velocity: | 595mps (1950fps) |
| Effective range: | c.300m (984ft) |

# M14

The M14 was developed in the USA during the 1950s to use the 7.62mm round then adopted as the standard NATO cartridge. It was effectively the Garand rifle with a 20-round box magazine and a selective fire facility. Yet the conversion was problematic and it took until 1957 before the production M14 appeared. Like the Garand, the M14 was reliable and hard-wearing, qualities it derived from a heavy investment in traditional methods of quality machining. This machining was more than needed to cope with the M14 when firing full auto, as the weight of the gun combined with the recoil forces of the 7.62mm round made it difficult to control. Indeed, the M14 was not really suited to automatic fire; the barrel would quickly overheat and could not be changed when hot. Despite this, the M14 was popular with US forces during the 1950s and 1960s. Production ended in 1964.

| | |
|---|---|
| Country of origin: | USA |
| Calibre: | 7.62mm NATO |
| Length: | 1117mm (44in) |
| Weight: | 3.88kg (8.55lb) |
| Barrel: | 558mm (22in), 4 grooves, rh |
| Feed/magazine capacity: | 20-round detachable box magazine |
| Operation: | Gas |
| Cyclic rate of fire: | 750rpm |
| Muzzle velocity: | 595mps (1950fps) |
| Effective range: | 800m (2600ft) plus |

# M21 Sniper Weapon System

**Although the 7.62mm M14 rifle was superseded by the new high-velocity 5.56mm guns, it lived on in several new incarnations. In particular, the US forces retained it as a sniping rifle (the 7.62mm round being more effective for sniping) designated the Rifle 7.62mm M14 National Match (Accurised), but which became known simply as the M21. The most obvious difference between the M21 and the standard M14 is the x3 Leatherwood Redfield telescopic sight. Less apparent are the improvements in the barrel, trigger and gas operation, which are manufactured to much higher standards and tolerances. The barrel is left without the usual chromium plating to avoid manufacturing errors. The M21 has proved to be a very capable sniper rifle and has been used outside of the USA by forces such as the Israeli Defence Force.**

| | |
|---|---|
| **Country of origin:** | USA |
| **Calibre:** | 7.62 x 51mm NATO |
| **Length:** | 1120mm (44.09in) |
| **Weight:** | 5.55kg (12.24lb) loaded |
| **Barrel:** | 559mm (22in), 4 grooves, rh |
| **Feed/magazine capacity:** | 20-round detachable box magazine |
| **Operation:** | Gas, self-loading |
| **Cyclic rate of fire:** | Semi-automatic |
| **Muzzle velocity:** | 853mps (2798fps) |
| **Effective range:** | 800m (2624ft) plus |

# Armalite AR-18

The AR-18 emerged from the Armalite factory in the mid-1960s in 5.56mm calibre, after a number of 7.62mm designs. The thinking behind the AR-18's design was to produce a simplified AR-15 (M16) for those countries which needed something cheap to buy, easy to maintain and simple to manufacture. Although some resemblances remain between the AR-18 and the AR-15, the former gun had a different piston-driven gas operation and a much broader use of metal stamping and pressing. The whole gun breaks down into only a few parts and both sights were on a single section so that the zero was kept even after disassembly and reassembly. The AR-18, however, failed because of M16 domination, despite the fact that some suggest that this accurate and powerful weapon was actually the superior of the two.

| | |
|---|---|
| Country of origin: | USA |
| Calibre: | 5.56mm M109 |
| Length: | 965mm (38in) stock extended; 730mm (28.74in) stock folded |
| Weight: | 3.04kg (6.70lb) |
| Barrel: | 463mm (18.25in), 4 grooves, rh |
| Feed/magazine capacity: | 20-round detachable box magazine |
| Operation: | Gas |
| Cyclic rate of fire: | 750rpm |
| Muzzle velocity: | 990mps (2530fps) |
| Effective range: | 500m (1640ft) plus |

# M16A1

**Designed by Eugene Stoner, the M16 first appeared as the 7.62mm AR-10 in the mid-1950s, followed by the AR-15, which was rechambered for the 5.56mm round. Licence to produce the AR-15 switched to Colt in 1959 and significant sales of the gun went out to Southeast Asia, the United Kingdom, the US Air Force and, finally, the US Army, when it was retitled the M16. Its early promise was almost completely undone during the early years of its service in Vietnam, where it had a tendency to jam in action and had to be kept very clean (not easy in the jungle). The problem was actually found to be a new propellant that caused excessive fouling; modifications to both gun and propellant left an excellent assault rifle, the M16A1. The plastic and pressed steel construction made it relatively light and the high-velocity of the round more than compensated for its small calibre in combat.**

| | |
|---|---|
| Country of origin: | USA |
| Calibre: | 5.56mm M193 |
| Length: | 990mm (39in) |
| Weight: | 2.86kg (6.3lb) |
| Barrel: | 508mm (20in), 6 grooves, rh |
| Feed/magazine capacity: | 30-round detachable box magazine |
| Operation: | Gas |
| Cyclic rate of fire: | 800rpm |
| Muzzle velocity: | 1000mps (3280fps) |
| Effective range: | 500m (1640ft) plus |

# Ruger Mini-14

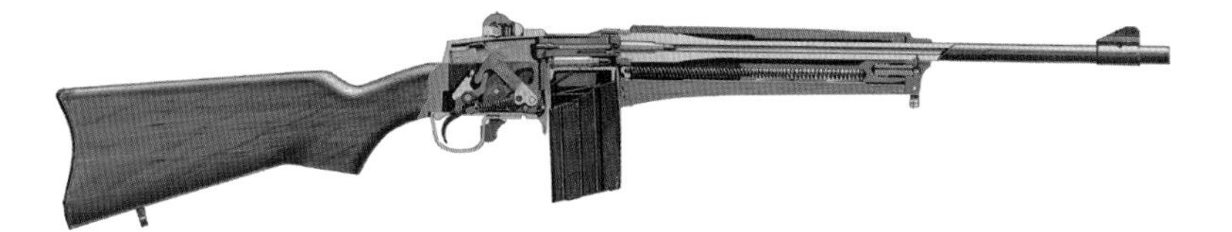

A quick glance at the Ruger Mini-14 shows that it is a more compact version of the M14 rifle, which in turn derives from the M1 Garand, the standard US rifle of World War II. Like both of those weapons, the Mini-14 works on a gas-operated rotating-bolt system, but it is calibrated for the high-velocity (but lighter weight) 5.56mm NATO or M193 rounds. Although the recoil of this round is less than the 7.62mm rounds of the M14, the gas pressure was still high enough to warrant some re-engineering. Thus, the gas piston is driven only a short distance via pressure on its cup head before the gas is vented out through an aperture and the piston carries the bolt back under its own momentum. The lightweight controllability of the Mini-14 has allowed its use by civilian, military and police customers, where it has proved popular.

| | |
|---|---|
| Country of origin: | USA |
| Calibre: | 5.56mm NATO or M193 |
| Length: | 946mm (37.24in) |
| Weight: | 2.9kg (6.70lb) |
| Barrel: | 470mm (18.5in), 6 grooves, rh |
| Feed/magazine capacity: | 5-, 10-, 20- or 30-round detachable box magazine |
| Operation: | Gas |
| Cyclic rate of fire: | 750rpm |
| Muzzle velocity: | 1005mps (3297fps) |
| Effective range: | 400m (1312ft) |

# Barret Light Fifty M82A1

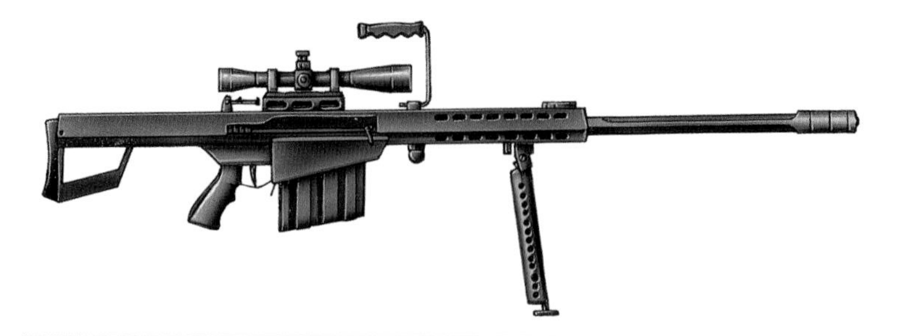

The Barrett Light Fifty M82A1 is a truly fearsome sniper weapon, firing a .50in Browning machine gun round over distances of up to and over a mile. The destructive force of the .50in round makes it decisive in both anti-personnel and anti-materiel roles. A semi-automatic, short-recoil weapon working off an 11-round box magazine, the Barrett controls its recoil mainly through a large muzzle brake which diverts some 30 per cent of its gases out at right angles to the direction of the barrel. The highly specialist nature of the weapon has somewhat limited its use and it is almost entirely in the hands of US special forces – although some have even turned up in terrorist use in Northern Ireland. It also saw service in the Gulf, where it proved a useful weapon over long ranges. It was, however, not the only .50in sniper rifle on the market and production of this particular weapon ceased in 1992.

| | |
|---|---|
| **Country of origin:** | USA |
| **Calibre:** | .50in Browning |
| **Length:** | 1549mm (60.98in) |
| **Weight:** | 14.7kg (32.41lb) |
| **Barrel:** | 838mm (33in), 8 grooves, rh |
| **Feed/magazine capacity:** | 11-round box magazine |
| **Operation:** | Short-recoil, semi-automatic |
| **Muzzle velocity:** | 843mps (2800fps) |
| **Effective range:** | 1000m (3280ft) plus |

# Iver Johnson Model 500

The Model 500 is a .50in calibre sniper rifle which developed out of the impressive Model 300. Why the Model 500 should have been developed when the Model 300 had serious knock-down power at 1500m (4921ft) is open to debate, but, whereas the Model 300 offered an acceptable performance, the Model 500 has been dogged by problems. The Model 500 is a truly advanced rifle, with a fluted and dampened heavy barrel and completely adjustable furniture. Such technologies, and the excellence of modern telescopic sights, have led the Model 500's manufacturers to claim an effective accuracy up to and over 2000m (6561ft). Yet it is unlikely that the Browning .50 machine-gun round could be controllable over such ranges and this, combined with the Model 500's enormous recoil and tendency to jam, has fettered the gun's commercial viability.

| | |
|---|---|
| **Country of origin:** | USA |
| **Calibre:** | .50in Browning |
| **Length:** | 1200mm (47.5in) |
| **Weight:** | 13.6kg (30lb) |
| **Barrel:** | 838mm (33in), rh |
| **Operation:** | Bolt action |
| **Muzzle velocity:** | 888mps (2914fps) |
| **Effective range:** | 2000m (6561ft) |

# M4 Carbine

Although carbine versions of the AR15/M16 family have been around since the 1970s, the M4 was only taken into US military service in the 1990s, essentially as a replacement for various pistols and submachine-guns that had fallen out of service. It has also been adopted by much of the US Army as a straightforward replacement for the much longer M16 rifles, being compact and easy to handle in urban warfare scenarios. It has a shortened barrel and a four-position sliding buttstock, the latter enabling the firearm to be adapted to a soldier's personal dimensions. In most other regards, including the 5.56 x 45mm calibre and the direct impingement gas-operating system, the weapon is the same as its larger brethren. Some firearms experts claim that the shorter barrel reduces muzzle velocity and so the efficacy of the 5.56mm bullet in terms of its take-down power.

| | |
|---|---|
| Country of origin: | USA |
| Calibre: | 5.56 x 45mm NATO |
| Length: | 840mm (33.07in) stock extended |
| Weight: | 2.79kg (6.15lb) |
| Barrel: | 370mm (14.57in) |
| Feed/magazine capacity: | 30-round detachable box magazine |
| Operation: | Gas |
| Cyclic rate of fire: | 700–950rpm |
| Muzzle velocity: | 884mps (2900fps) |
| Effective range: | 600m (1969ft) |

# M14 Enhanced Battle Rifle (EBR)

Developed directly from the M14 sniper rifle, progenitor of the M21 sniper rifle, the M14 Enhanced Battle Rifle (EBR) is issued to US special forces. Adapted to provide infantry squadrons with the firepower of a semi-automatic weapon but with the range and accuracy of a sniper rifle, the M14 EBR uses the same gas-operated/rotating bolt action and air-cooled barrels as the original M14. It differs from the standard M14 by having an adjustable telescopic buttstock, an M4-style pistol grip and four Picatinny accessory rails. It can also carry two different scopes at the same time. In addition, it can be converted back to the standard M14 if necessary. It has been in service with the US NAVY SEALs since 2004 and is being introduced into select infantry squadrons of the US Army. It also exists in a civilian version.

| | |
|---|---|
| Country of origin: | USA |
| Calibre: | 7.62mm (.3in) NATO |
| Length: | 889mm (35in) |
| Weight: | 5.1kg (11.24lb) |
| Barrel: | 457mm (18in) |
| Feed/magazine capacity: | 10- or 20-round detachable box magazine |
| Operation: | Gas |
| Cyclic rate of fire: | 725rpm |
| Muzzle velocity: | 975.4mps (3200fps) |
| Effective range: | 800m (2624ft) + |

# M110 SASS

The M110 SASS is a new generation of 7.62 x 51mm NATO sniper rifle, developed on the basis of a semi-automatic, gas-operated, rotating-bolt design. Fed from 10- or 20-round box magazines, the semi-auto capability offers the shooter defensive firepower when necessary, but the long-range accuracy is delivered by a heavy freefloating barrel. It is capable of putting a group of shots within 28mm (1.10in) at 91.4m (300ft), and reports back from combat testing in Afghanistan have been extremely positive. Standard fitment on the rifle includes a folding bipod, a collapsible buttstock and the XM151 3.5–10 x variable scope on the MIL-STD-1913 rail system. The barrel also has the option of fitting with a suppressor. Relatively light for a sniper rifle at 7kg (15.43lb), the M110 was voted the second best product of 2007 by the US Army.

| | |
|---|---|
| Country of origin: | USA |
| Calibre: | 7.62 x 51mm NATO |
| Length: | 1028mm (40.47in) |
| Weight: | 7kg (15.43lb) |
| Barrel: | 508mm (20in) |
| Feed/magazine capacity: | 10- or 20-round detachable box magazine |
| Operation: | Gas |
| Cyclic rate of fire: | Not applicable |
| Muzzle velocity: | 784mps (2571fps) |
| Effective range: | 800m (2624ft) |

# FN SCAR

The FN SCAR was designed by FN USA to meet the US Army's requirement for a new Special Operations Forces assault rifle. (The acronym stands for SOF Combat Assault Rifle.) With deliveries beginning in 2005, the SCAR is a typical high-quality FN firearm, and constitutes a completely new design rather than a redevelopment of an existing model. SCAR is a modular series rather than a single weapon. The core of the rifles remains a gas-operated, rotating-bolt system that uses a short-stroke piston movement for efficient weapon cycling. The stock is adjustable for comb height and length of pull, and the guns can take various optical sights or rely on folding, adjustable iron sights. The weapon configuration changes according to what barrel is fitted: free-floating hammer-forged barrels in 254, 356 and 457mm (10, 14 and 18in) lengths are available, each fitted with a three-pronged flash suppressor.

| | |
|---|---|
| Country of origin: | USA |
| Calibre: | 5.56 x 45mm NATO |
| Length: | 737–990mm (29.02–38.98in) |
| Weight: | 3.5kg (7.72lb) |
| Barrel: | 457mm (17.99in) |
| Feed/magazine capacity: | 30-round detachable box magazine |
| Operation: | Gas |
| Cyclic rate of fire: | 600pm |
| Muzzle velocity: | 910mps (2986fps) |
| Effective range: | 600m (1969ft) |

# Simonov AVS-36

The Soviet Army's experience of automatic rifles began somewhat inauspiciously when, in the 1930s, they adopted the AVS-36 designed by Sergei Gavrilovich Simonov (although it is predated by Federov's Avtomat) without ensuring its performance through proper service trials. Like many later Soviet assault rifles, the gas piston was set above the barrel, yet its operation struggled to contain the power of the Soviet 7.62mm round and it was also susceptible to the ingress of dirt. These factors, combined with a hefty recoil and muzzle blast, curtailed the future of the AVS-36 and, by 1938, it was out of production, if not actually out of service. Its replacement was the Tokarev SVT, but Simonov did go on to design the quality SKS rifle which used the new M193 7.62mm intermediate round and saw service after the end of World War II.

| | |
|---|---|
| Country of origin: | USSR/Russia |
| Calibre: | 7.62 x 52R |
| Length: | 1260mm (49.6in) |
| Weight: | 4.4kg (9.7lb) |
| Barrel: | 627mm (24.69in), 4 grooves, rh |
| Feed/magazine capacity: | 15-round detachable box magazine |
| Operation: | Gas |
| Cyclic rate of fire: | 600rpm |
| Muzzle velocity: | 835mps (2740fps) |
| Effective range: | 500m (1640ft) plus |

# Tokarev SVT-40

Feydor Vassilivich Tokarev not only produced a famous Soviet pistol (the TT-33), but also a fairly well designed gas-operated assault rifle, the Samozaryadnaya Vintovka obr 1938, or SVT-38. It was a lengthy gun with a simple locking action consisting of a block which was cammed downwards into a recess in the receiver floor, this being released by the backward motion of the bolt carrier. The SVT-38 had a heavy muzzle blast, but Tokarev controlled this through a six-baffle muzzle brake. The gun was superseded in 1940 by the SVT-40 and this became the dominant model, with some two million produced between 1939 and 1945. Visually, the two guns were very similar. The SVT-40 had its cleaning rod situated under the barrel, rather than against the side of the forestock. Also, the SVT-40's muzzle brake came in six- or two-baffle versions. Everything else stayed almost exactly the same.

| | |
|---|---|
| Country of origin: | USSR/Russia |
| Calibre: | 7.62 x 52R |
| Length: | 1226mm (48.27in) |
| Weight: | 3.90kg (8.6lb) |
| Barrel: | 610mm (25in), 4 grooves, rh |
| Feed/magazine capacity: | 10-round detachable box magazine |
| Operation: | Gas |
| Cyclic rate of fire: | Semi-automatic |
| Muzzle velocity: | 840mps (2755fps) |
| Effective range: | 500m (1640ft) plus |

# Simonov SKS

Simonov's SKS carbine was a Soviet attempt to find a weapon suitable for firing the 7.62mm Short (Kurz) cartridge discovered in captured German automatic MP44s. Of course, in later years the AK-47 became the defining Soviet weapon for the intermediate round, but the SKS came first, being trialled in 1944 and produced from 1946. A simple, robust, if rather heavy, weapon, the SKS was a solid gun which, although soon relegated to ceremonial use within the Soviet Union, spread around the world through other communist armies within Europe and Asia. As a single-shot weapon, it could put out fire to about 400m (1312ft), an especially useful combat range. Although features such as the folding bayonet under the muzzle now appear anachronistic to the modern eye, the SKS will no doubt keep making appearances for many years to come.

| | |
|---|---|
| Country of origin: | USSR/Russia |
| Calibre: | 7.62mm Soviet M1943 |
| Length: | 1022mm (40.2in) stock extended |
| Weight: | 3.86kg (8.51lb) |
| Barrel: | 520mm (20.47in), 4 grooves, rh |
| Feed/magazine capacity: | 10-round detachable box magazine |
| Operation: | Gas |
| Cyclic rate of fire: | Single shot |
| Muzzle velocity: | 735mps (2410fps) |
| Effective range: | 500m (1312ft) |

# Kalashnikov AK-47

The AK firearms are the most produced and distributed series of small arms in history. The series started with the AK-47, developed just after World War II to provide an intermediate-range infantry weapon which was resilient and fast-firing. Part of Mikhail Kalashnikov's inspiration was the German MP44 and its use of the new 7.92mm Kurz cartridge. Although Simonov produced a weapon to the new specification which went into production before the AK-47, it was the latter that met with incredible success. The AK-47 was a simple gas-operated design using a rotating bolt. It had a chromium-plated barrel and generally high-quality machining and finishing. It took until 1959 to perfect the design and production processes, but once this had been achieved, the Soviets were left with a robust, easy-to-produce firearm. It has spread around the globe.

| | |
|---|---|
| Country of origin: | USSR/Russia |
| Calibre: | 7.62mm Soviet M1943 |
| Length: | 880mm (34.65in) |
| Weight: | 4.3kg (9.48lb) |
| Barrel: | 415mm (16.34in), 4 grooves, rh |
| Feed/magazine capacity: | 30-round detachable box magazine |
| Operation: | Gas |
| Cyclic rate of fire: | 600rpm |
| Muzzle velocity: | 600mps (2350fps) |
| Effective range: | 400m (1312ft) |

# Kalashnikov AKM

While the original AK-47 was a superb gun in most respects, there were problems in the quality of its stamped steel receiver which let down its potentially good reliability. In 1951, the switch was made to a machined receiver, but this then made the gun's production cost per unit rise considerably. The solution was found in 1959 in the AKM (the M stands for 'Modernised'), the most prolific of the AK series and the one most likely to be encountered when Kalashnikovs are present on the battlefield. It featured a higher quality stamped receiver ideal for the gun and it also took on several other minor improvements. These included an angled muzzle which acted as a basic compensator to control muzzle climb and a newly designed bayonet which could convert into a wire cutter. An AKM can usually be distinguished from an AK-47 by the recess above the magazine housing.

| | |
|---|---|
| Country of origin: | USSR/Russia |
| Calibre: | 7.62mm Soviet M1943 |
| Length: | 880mm (34.65in) |
| Weight: | 4.3kg (9.48lb) |
| Barrel: | 415mm (16.34in), 4 grooves, rh |
| Feed/magazine capacity: | 30-round detachable box magazine |
| Operation: | Gas |
| Cyclic rate of fire: | 600rpm |
| Muzzle velocity: | 600mps (2350fps) |
| Effective range: | 400m (1312ft) |

# Kalashnikov AK-74

The AK-74 is now the standard rifle of the Russian armed forces and it began its replacement of the AKM rifle during the late 1970s. To all intents and purposes, it is the 7.62mm AKM rifle rechambered and modified for the smaller 5.45mm cartridge, part of the worldwide trend for smaller calibre, high-velocity weapons which occurred during the 1960s and 1970s – although the West opted for the 5.56mm round. To compensate for the small calibre, the round itself is steel cored, hollow tip and with a rearward centre of gravity, the result being that the bullet tumbles through the target on impact, causing far greater damage to a body than a 'clean' bullet entry – an effect outlawed by many nations. The distinguishing feature of the AK-74 is the large muzzle brake which, combined with the round, means that the gun produces almost no recoil.

| | |
|---|---|
| Country of origin: | USSR/Russia |
| Calibre: | 5.45mm M74 |
| Length: | 940mm stock extended |
| Weight: | 3.6kg (7.94lb) |
| Barrel: | 400mm (15.8in), 4 grooves, rh |
| Feed/magazine capacity: | 30-round box |
| Operation: | Gas |
| Cyclic rate of fire: | 650rpm |
| Muzzle velocity: | 900mps (2952fps) |
| Effective range: | 300m (1000ft) |

# Kalashnikov AKS-74

Issued at the same time as its fixed stock counterpart, the AKS-74 is little different from the standard AK-74 rifle. In fact the only change is the use of the tubular folding stock. This folds to the left of the rifle's body, in contrast to the earlier AK-series folding stock models, whose stock would sit under the rifle when not in use. Another version of the rifle, the AK-74M, is also fitted with a folding stock, but this is a solid plastic version which folds to the right of the receiver. All the AK-74 series rifles are fitted with the recoil-reducing muzzle brake, which follows the Soviet doctrine of preferring troops to produce grouped automatic suppressive fire rather than aimed shots. One of the drawbacks of the muzzle brake, however, is that it does not reduce muzzle flash, which is approximately three times normal. The AK-74 series is licence-produced in Bulgaria, Hungary, Poland and Romania.

| | |
|---|---|
| Country of origin: | USSR/Russia |
| Calibre: | 5.45mm M74 |
| Length: | 690mm (27.2in) stock folded |
| Weight: | 3.6kg (7.94lb) |
| Barrel: | 400mm (15.8in), 4 grooves, rh |
| Feed/magazine capacity: | 30-round box |
| Operation: | Gas |
| Cyclic rate of fire: | 650rpm |
| Muzzle velocity: | 900mps (2952fps) |
| Effective range: | 300m (1000ft) |

# AK 103

Though Kalashnikov produces small-calibre rifles (5.45mm/5.56mm) in line with the world-wide trend towards calibre reduction, it still manufactures a 7.62mm AK-series rifle in the AK 103. This rifle balances with the 5.56mm AK 101, which tends to be seen in nations which have large stockpiles of Western NATO ammunition. By contrast the AK 103 is much more prominent in former Soviet allies and within Russia itself, as it fires Soviet-era ammunition. The AK 103 has all the reliability and strengths of Kalashnikov rifles in general. It features an extensive use of plastic in its construction: the butt (which is folding), foregrip (when fitted), handguard, pistolgrip and magazine are all produced in plastic, thus keeping the weight down to 3.4kg (7.49lb). The AK 103 ranks alongside the AK 74 as a standard Russian assault rifle.

| | |
|---|---|
| Country of origin: | USSR/Russia |
| Calibre: | 7.62 x 39mm |
| Length: | 943mm (37.1in) |
| Weight: | 3.4kg (7.49lb) |
| Barrel: | 415mm (16.3in) |
| Feed/magazine capacity: | 30-round detachable box magazine |
| Operation: | Gas operated |
| Cyclic rate of fire: | 600rpm |
| Muzzle velocity: | Not known |
| Effective range: | 300m (984ft) plus |

# Dragunov SVD

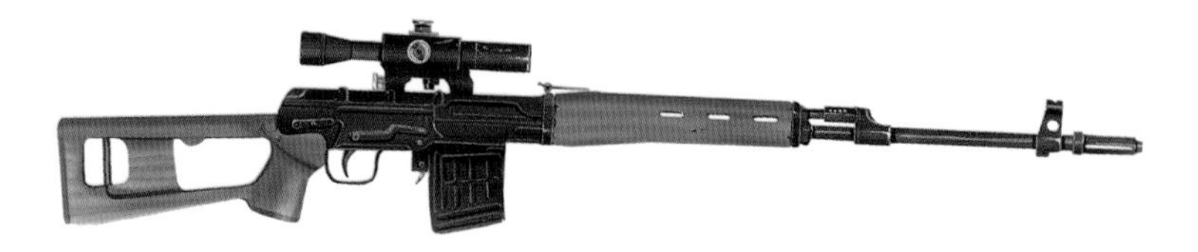

The Dragunov bears a family resemblance to the AK series of weapons, but it has been extensively re-engineered to meet the demands of the sniping role. It is semi-automatic, but the gas system works with a short-stroke piston, rather than the AK's long stroke to improve the weapon's stability. Accuracy comes from a combination of a very long barrel and the standard fitment of a PSO-1 x4 telescopic sight which can also act upon infra-red detections to give a night-sight capability. It has a first shot range of around 1000m (3280ft) and it is able to achieve this range by swapping the AK's 7.62mm Short cartridge for the older rimmed 7.62 x 54R cartridge, the ancestry of which stretches back to the late 19th century. The overall result is a very capable sniper weapon which is both precise and rugged, and it has served well since it was introduced in 1963.

| | |
|---|---|
| Country of origin: | USSR/Russia |
| Calibre: | 7.62mm x 54R Soviet |
| Length: | 1225mm (48.20in) |
| Weight: | 4.31kg (9.5lb) |
| Barrel: | 610mm (24in), 4 grooves, rh |
| Feed/magazine capacity: | 10-round detachable box magazine |
| Operation: | Gas |
| Cyclic rate of fire: | N/A |
| Muzzle velocity: | 828mps (2720fps) |
| Effective range: | 1000m (3280ft) |

# AN-94 Abakan

The AN-94 'Abakan' was developed during the late 1980s and early 1990s as a replacement for the 5.45 x 39mm AN-74 series of rifles in Russian service. It is a technologically complex and innovative weapon. Although gas-operated, it uses a system labelled 'blowback shifted pulse', which is capable of delivering a two-round burst (the rifle's standard burst setting) at a rate of 1800rpm, the user essentially feeling one recoil impulse for the two shots, thereby improving accuracy. When fired on full-auto, the first two shots are at this higher rate, before settling down to a cyclical rate of 600rpm. The cost of this ability is a complicated internal mechanism, which sacrifices the simplicity of the traditional AK model. The Abakan can take standard 30- or 45-round AK-74 magazines or new 60-round boxes, and includes an advanced five-aperture rear sighting system.

| | |
|---|---|
| Country of origin: | Russia/CIS |
| Calibre: | 5.45 x 39mm |
| Length: | 943mm (37.13in) |
| Weight: | 3.85kg (8.49lb) |
| Barrel: | 520mm (20.47in) |
| Feed/magazine capacity: | 30-round detachable box magazine |
| Operation: | gas, rotating bolt |
| Cyclic rate of fire: | 600rpm |
| Muzzle velocity: | not available |
| Effective range: | 500m (1640ft) |

# AK-107

The AK-107 was developed as a cheaper alternative to the AN-94. It is capable of full- or semi automatic fire as well as three-round bursts. Noted for its Balanced Automatic Recoil System (BARS), the AK-107 balances the gas operation of the firing operation and reduces the recoil. Rather than the wood and metal used for the AK-47, the AK-107 is made of lightweight fibreglass-reinforced polymers which are also cheaper to produce. The shoulder stock is hinged, so the gun can be compacted to a more manageable length. The gun can mount a variety of available optics, aiming accessories and the 40mm GP-25 series single-shot, underslung grenade launcher. Geared towards a more Western market, the AK-108 is a derivative of the AK-107 chambered for the 5.56 x 45mm NATO instead of the the 5.45 x 39mm Soviet cartridge.

| | |
|---|---|
| Country of origin: | Russia |
| Calibre: | 5.45mm (.21in) |
| Length: | 943mm (37.1in) |
| Weight: | 3.8kg (8.38lb) |
| Barrel: | 415mm (16.3in) |
| Feed/magazine capacity: | 30-round detachable box magazine |
| Operation: | Gas |
| Cyclic rate of fire: | 850rpm |
| Muzzle velocity: | 900m/sec (2,953 ft/sec) |
| Effective range: | 500m (1640ft) |

# AK-200

Prototypes were unveiled in 2010 of the AK-200, the latest in the AK series of Russian assault rifles which began with the AK-47. Production of the AK-200 started in 2011 and once in service the gun will be designated the AK-12. The rifle reflects modern preferences for accessory rails and advanced materials, allowing the attachment of modular equipment, including advanced optical sight combinations, laser illuminators, flashlights, vertical foregrips, bipods and grenade launchers. As with many of today's sniper rifles, the shoulder stock is adjustable to suit the firer's requirements. In terms of design, the AK-200 resembles the AK-47, but without the use of wood. Instead, the AK-200 will use composites and plastics to keep its operating weight at a minimum but its recoil forces in check. As illustrated, it includes optional foregrips.

| | |
|---|---|
| Country of origin: | Russia |
| Calibre: | 5.56mm (0.21in) |
| Length: | 943mm (37.1in) |
| Weight: | 3.8 kg (8.38 lb) |
| Barrel: | 415mm (16.3in) |
| Feed/magazine capacity: | 30-round detachable box magazine |
| Operation: | Gas |
| Cyclic rate of fire: | 850rpm |
| Muzzle velocity: | 900mps (2953fps) |
| Effective range: | 500m (1640ft) |

# Franchi SPAS Model 12

The Franchi SPAS 12 is a purpose-designed combat shotgun and boasts many features which make it eminently versatile in roles from crowd control to military use. It first emerged as the Model 11 in 1979, and the Model 12 followed shortly afterwards with a folding version of the skeleton metal stock which also featured an arm hook for one-armed firing. Both models are pump action and semi-automatic – the mode is selected by a single button under the fore-end. They can even fire full auto, shooting out everything from bird shot and single slugs to tear-gas projectiles. Its pattern of spread using a standard 12-gauge cartridge is 0.9m (35.4in) at 40m (131ft), making it a fearsome weapon to face. The SPAS Models 11 and 12 have revolutionised combat shotgun design, and the black phosphated metal creates a sense of visual threat which alone has proved useful in action.

| | |
|---|---|
| **Country of origin:** | Italy |
| **Gauge/calibre:** | 12 |
| **Length:** | 930mm (36.6in) stock extended; 710mm (27.95in) stock folded |
| **Weight:** | 4.2kg (9.26lb) |
| **Barrel:** | 460mm (18.11in) |
| **Feed/magazine capacity:** | 7-round integral tubular magazine |
| **Operation:** | Pump action and gas (for semi-auto/full-auto modes) |
| **Rate of fire:** | Cyclic 240rpm when on full-auto |
| **Muzzle velocity:** | Variable, depending on type of ammunition |
| **Effective range:** | 100m (328ft) |

# Franchi SPAS 15

One of the problems of the conventional pump-action shotgun is the time it takes to reload the weapon once all the shells have been emptied from the tubular magazine. After considering this problem, the Franchi concern developed several magazine-loading prototypes, including a bullpup configuration weapon which did not make it into production. There followed the SPAS 14, a magazine-loaded version of the SPAS 12, and then the SPAS 15. In appearance, the SPAS 15 looks much like an assault rifle, with a detachable 10-round magazine and a fire selection facility allowing the user to switch between semi-automatic and pump action. Full automatic was considered by Franchi, but the ergonomic and practical problems of excessive recoil were never overcome and so this is not yet a feature offered by the company's shotguns.

| | |
|---|---|
| Country of origin: | Italy |
| Gauge/Calibre: | 12 |
| Length: | 980 or 1000mm (38.58 or 39.3in) |
| Weight: | 3.9 or 4.1kg (8.5 or 9lb) |
| Barrel: | 450mm (17.71in) |
| Feed/magazine capacity: | 10-round detachable box magazine |
| Operation: | Pump action and gas |
| Rate of fire: | Semi-automatic |
| Muzzle velocity: | Variable, depending on type of ammunition |
| Effective range: | 100m (328ft) |

# Remington M870

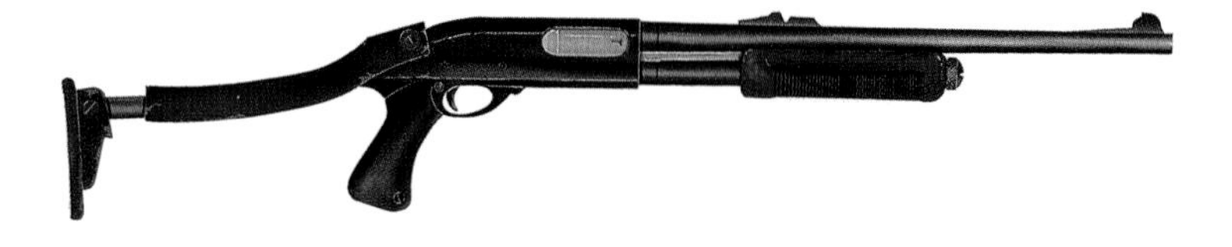

Remington has had a long history of firearms manufacture in all weapon types, not least shotguns. Its M870 series of guns has been extensively used for police and military work across the world, particularly for riot and close-quarter operations, respectively. Perhaps the defining military model was the M870 Mk 1. This was adopted by the US Marines Corps in the mid-1960s. It was a standard pump-action shotgun that was durable, powerful, had a good magazine capacity (seven rounds) and was decisive in putting a man down at close range. This killing force was tested in anger in Vietnam by Marine and US Navy SEAL teams. It could fire standard shot, flechettes, solid slugs or various other forms of ammunition. The Mk 1 still serves today around the world in assorted roles, alongside a broad family of other M870 weapons.

| | |
|---|---|
| Country of origin: | USA |
| Gauge/Calibre: | 12 gauge |
| Length: | 1060mm (41.73in) |
| Weight: | 3.6kg (7.94lb) |
| Barrel: | 533mm (21in) |
| Feed/magazine capacity: | 7-round integral tubular magazine |
| Operation: | Pump action |
| Muzzle velocity: | Variable, depending on type of ammunition |
| Effective range: | 100m (328ft) |

# Ithaca Model 37 M and P

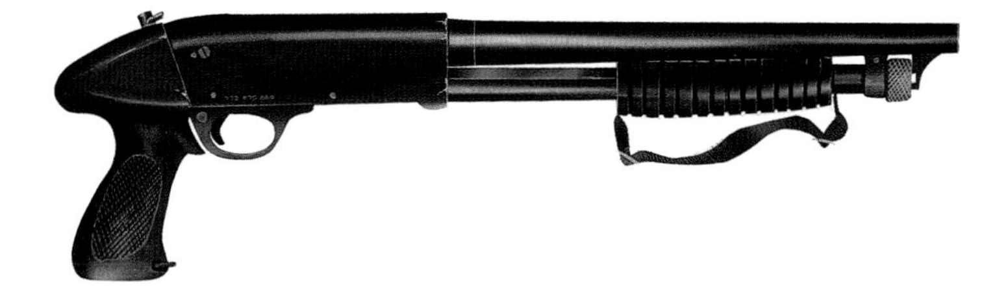

The Ithaca 37 series of shotguns is not a range of commercial guns that have simply stepped into military and police use – they are purpose-designed for these specialist roles. Ithaca 37 guns served in World War II and have kept their popularity, even, it is rumoured, amongst special forces soldiers. Most prevalent today is the Ithaca 37 M and P ('Military and Police'), a 12-gauge shotgun which comes in two barrel lengths – 470mm (18.5in) and 508mm (20in) – and has either a five- or an eight-round tubular magazine. Another model, the charmingly-named DS ('Deerslayer'), has similar specifications, but has a barrel engineered to fire heavy slugs when necessary. The DS led to the Model LAPD, a special gun with modified sights and furniture, and improved carrying straps and sling swivels for specific police use.

| | |
|---|---|
| Country of origin: | USA |
| Gauge/calibre: | 12 |
| Length: | 1016mm (40in) for 508mm barrel |
| Weight: | 2.94kg (6.48lb) or 3.06kg (6.75lb) |
| Barrel: | 470mm (18.5in) or 508mm (20in) |
| Feed/magazine capacity: | 5- or 8-round integral tubular magazine |
| Operation: | Pump action |
| Muzzle velocity: | Variable, depending on type of ammunition |
| Effective range: | 100m (328ft) |

# Mossberg ATPS 500

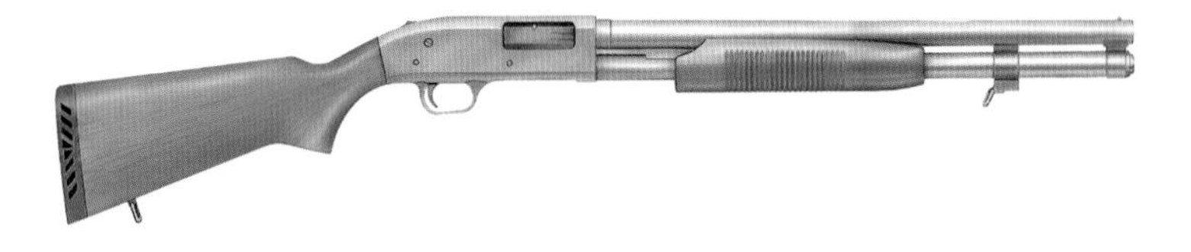

The Mossberg 500 series of shotguns began its life in the 1960s as a new exploration into combat shotguns for the company O.F. Mossberg & Sons Inc. Its current forms include models for hunting, home security and military use, but the ATPS was specifically designed for military applications. The Model 500 is in most senses a conventional pump-action shotgun, even if a very robust one. The quality of machining is outstanding and all the internal components are made from very high-grade steel. Special features of the ATPS, which reflect its possible usage in combat, include a bayonet mount and even a telescopic sight mounting for greater accuracy when firing slugs. A new bullpup-type Model 500 known as the Bullpup 12 is also now on the market, but the conventional 500 retains great loyalty amongst its many users.

| | |
|---|---|
| Country of origin: | USA |
| Gauge/calibre: | 12 |
| Length: | 1070mm (42in) |
| Weight: | 3.3kg (7.27lb) |
| Barrel: | 510mm (20in) |
| Feed/magazine capacity: | 6-round integral tubular magazine |
| Operation: | Pump action |
| Muzzle velocity: | Variable, depending on type of ammunition |
| Effective range: | 100m (328ft) |

# Pancor Jackhammer

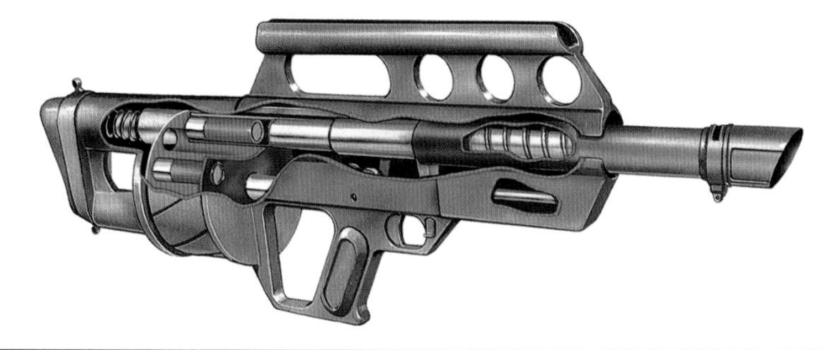

The space-age appearance of the Pancor Jackhammer denotes its place amongst a new breed of combat shotguns. It is set in a bullpup configuration with the 10-round 'ammunition cassette' set behind the trigger group. This cassette is a revolver-feed magazine which, once emptied, is detached and replaced by a new cassette ready to fire – the old cassette is not reloaded. Neither is there any ejection of rounds, they are simply retained inside the cassette. Each cassette comes pre-loaded in sealed packaging which denotes the type of ammunition within. The Jackhammer has a gas-operated system which, on full-automatic setting, is capable of a cyclic rate of fire of 240rpm. Unlike many other automatic shotguns, the Jackhammer can sustain this fire owing to an effective muzzle compensator and a very rugged construction in steel and high-impact plastics.

| | |
|---|---|
| Country of origin: | USA |
| Gauge/calibre: | 12 gauge |
| Length: | 762mm (30in) |
| Weight: | 4.57kg (10lb) loaded |
| Barrel: | 457mm (18in) |
| Feed/magazine capacity: | 10-round detachable pre-loaded rotary cassette |
| Operation: | Gas |
| Muzzle velocity: | Variable, depending on type of ammunition |
| Effective range: | 200m (656ft) plus |

# Winchester 12 Defender

Winchester is one of the USA's most venerable rifle manufacturers and it produces a high-quality range of shotguns. The Winchester 12, a 12-gauge pump-action shotgun which entered service in World War II, is still used today for police and military work. The basic specifications of the Winchester 12 include a six- or seven-round under-barrel tubular magazine (the number of rounds carried depends on the type of cartridges), various sight configurations for the firing of shot or slugs, and good-quality finishes to the metalwork which are either blued, Parkerised or stainless steel (the latter being a particular police version). The Defender Model is the essential Model 12 and current issues can be with a stock in conventional format or without a stock and featuring a pistol grip, these being ideal for security operations where the weapon needs to be concealed until required.

| | |
|---|---|
| Country of origin: | USA |
| Gauge/calibre: | 12 gauge |
| Length: | Not available |
| Weight: | 3.06kg (6.74lb), variations apply |
| Barrel: | 457mm (18in) |
| Feed/magazine capacity: | 6- or 7-round integral tubular magazine |
| Operation: | Pump action |
| Muzzle velocity: | Variable, depending on type of ammunition |
| Effective range: | 200m (656ft) |

# Granatpistole

The Granatpistole comes from the estimable name of Heckler & Koch and is a fairly conventional break-open 40mm (1.57in) grenade launcher. Its primary virtues are its portability and size, as the stock folds down and gives the weapon a convenient 463mm (18.2in) length. The fact that the Granatpistole can also be safely fired in the stock-folded position has added to its overall credibility as a weapon, and it has gone into service successfully with the German army and also with various security and special forces troops. Like most grenade launchers, the Granatpistole can fire a variety of shells, such as high-explosive and tear gas, and its range of about 300m (984ft) means that it has practical applications both for long-range lob shots or flat-trajectory firing at closer ranges. It is likely to remain in service for some time.

| | |
|---|---|
| Country of origin: | Germany |
| Calibre: | 40mm |
| Length: | 683mm (27in) stock extended; 463mm (18.2in) stock folded |
| Weight: | 2.3kg (5lb) |
| Barrel: | Not available |
| Feed/magazine capacity: | Single round, breech-loaded |
| Operation: | Breech-loaded |
| Muzzle velocity: | 75mps (245fps) |
| Effective range: | 350m (1148ft) |

# M79 Grenade Launcher

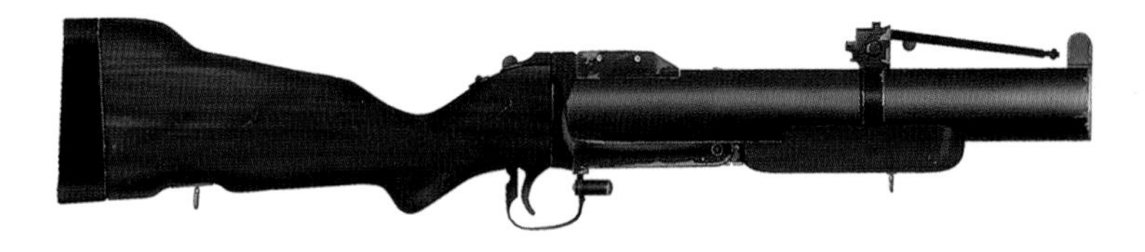

The M79 was a breech-loading single-shot grenade launcher which became a familiar weapon during the Vietnam conflict. It fired a 40mm (1.57in) grenade to a range of about 400m (1312ft) and, when loaded with high explosive, had an impressive kill radius of more than 5m (16ft). A variety of shells could be fired from the M79, including buckshot, flechette, airburst, smoke, flares and CS gas canisters, and the rifled barrel ensured that the M79 had good accuracy over an effective range of around 150m (490ft). The spin produced by the rifling also caused the grenade to arm itself after about 30m (98ft) of flight through a shift in internal weights. Some 350,000 'Bloopers' (as they were known by the troops in Vietnam) were produced between 1961 and 1971, and the M79 can still be found in use across the world today.

| | |
|---|---|
| Country of origin: | USA |
| Calibre: | 40mm |
| Length: | 783mm (29in) |
| Weight: | 2.95kg (6.5lb) loaded |
| Barrel: | Not known |
| Feed/magazine capacity: | Single round |
| Operation: | Breech-loaded, single shot |
| Muzzle velocity: | 75mps (245fps) |
| Effective range: | 150m (492ft) |

# M203 Grenade Launcher

The M203 was the replacement for the M79 Grenade Launcher and came into service in 1970, after combat trials in Vietnam. It was a development of the AAI Corporation and offered several advantages over its predecessor. Not least of these was that it could be fitted beneath a standard M16 rifle, but not interfere with the operation of the rifle itself, thus giving the operator a means of defending himself. The M203 is breech-loaded with a single 40mm (1.57in) grenade with ejection provided by a pump-action system. Its range of up to 400m (1312ft) far outreached the M79 and the M203 had a fairly high degree of accuracy within that range. Its grenades offered a standard range of anti-personnel and anti-vehicle warheads, and a later stand-alone version fitted with its own stock and sights proved a capable riot control weapon for firing tear gas canisters.

| | |
|---|---|
| Country of origin: | USA |
| Calibre: | 40mm |
| Length: | 380mm (15in) |
| Weight: | 1.63kg (3.5lb) loaded |
| Barrel: | Not available |
| Feed/magazine capacity: | Single round, breech-loaded |
| Operation: | Breech-loaded, single shot, pump action |
| Muzzle velocity: | 75mps (245fps) |
| Effective range: | 400m (1312ft) |

# Mark 19 Grenade Launcher

The Mark 19 Grenade Launcher became a popular harassing tool against the Vietcong during the Vietnam War. There it was first mounted onto a number of US river patrol craft used to control the Vietnamese coastline, which could bombard enemy positions ranged along the river bank with a stream of high-explosive 40mm (1.57in) grenades. This proved to be an effective application for the weapon and the Mark 19's customers increased to include the US Army and also Israeli forces, with both finding the Mark 19 a useful and deadly weapon when placed on vehicle mountings. A blowback grenade launcher capable of full-automatic fire feeding from a disintegrating link belt-feed, it fires from an open bolt (the bolt stays back from the chamber between firing), which helps the weapon to stay cool when firing repeatedly without pause.

| | |
|---|---|
| Country of origin: | USA |
| Calibre: | 40mm |
| Length: | 1028mm (40in) |
| Weight: | 34kg (74.8lb) |
| Barrel: | Not available |
| Feed/magazine capacity: | Belt feed, disintegrating-link belt |
| Operation: | Blowback |
| Muzzle velocity: | 240mps (790fps) |
| Effective range: | 1600m (5249ft) |

# AGS-17

The AGS-17 is one of a series of belt-fed automatic grenade launchers which emerged in various world armies from the 1960s onwards. Unlike the similar US 40mm (1.57in) Mark 19, which it resembled, the AGS-17 came in 30mm (1.18in) calibre – although it did also operate on the basis of a blowback action. The recoil forces of this operation were controlled through the mount, and the AGS-17 could be either used as an infantry weapon or operated from a helicopter or vehicle platform. The feed system used on the gun was belt feed, and this enabled the operator to lay down a systematic bombardment on targets to ranges of up to 1200m (3927ft). In this capacity, it was used to lethal effect against the guerrilla fighters in Afghanistan's mountainous terrain during the Soviet occupation of the country in the 1980s.

| | |
|---|---|
| Country of origin: | USSR/Russia |
| Calibre: | 30mm |
| Length: | 840mm (33in) |
| Weight: | 18kg (39.6lb) without tripod |
| Barrel: | Not available |
| Feed/magazine capacity: | Belt feed |
| Operation: | Blowback, automatic |
| Muzzle velocity: | Not known |
| Effective range: | 1200m (3927ft) |

# Milcor MGL

**The South African Milcor MGL is one of a new breed of revolver-type grenade launchers designed for roles ranging from hand-held riot-control applications to airborne and vehicle use in combat. Its semi-automatic operation allows six 40mm grenades to be fired in less than three seconds, giving it impressive firepower, and the ammunition types fired include high explosive, plastic shot, smoke, tear gas and baton rounds (the ammunition types can be mixed in the magazine if required). In combat the MGL's effective range moves from a minimum of 30m (98ft) to a maximum of 400m (1312ft), and accuracy is sustained by the occluded eye gunsight (OEG) featuring a built-in range estimator. The MGL is a gas-operated weapon and is also available in a twin mounted configuration for use on vehicles or on its own tripod fitting.**

| | |
|---|---|
| **Country of Origin:** | South Africa |
| **Calibre:** | 40mm |
| **Length:** | 788mm (31in) stock extended; 566mm (22.2in) stock folded |
| **Weight:** | 5.3kg (11.6lb) |
| **Barrel:** | 310mm (12.2in) 6 grooves, rh |
| **Operation:** | Gas, semi-automatic |
| **Feed/Magazine capacity:** | 6-round revolving cylinder |
| **Muzzle Velocity:** | 75mps (245fps) |
| **Effective Range:** | 400m (1312ft) |

# Brunswick RAW

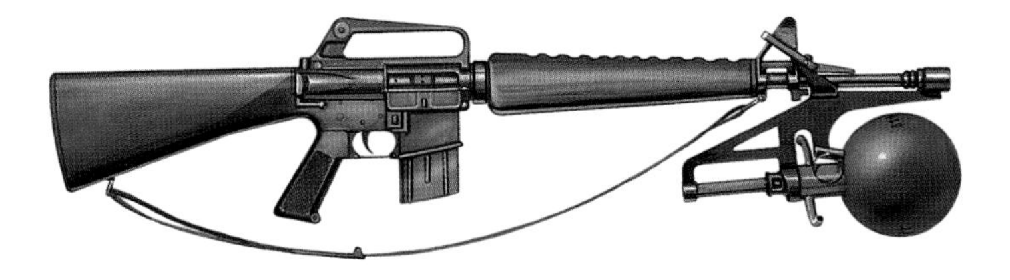

While the M203 grenade launcher has become a common mount on the US M16 rifle, the Brunswick Rifleman's Assault Weapon (RAW) still has some way to go to acceptance. This unusual spherical bomb is fired from the M16 and is intended for use against light vehicles and in urban settings. In flight, the bomb is rocket assisted; when it strikes the target, its contents – 1.27kg (2.7lb) of high explosive – are plastered onto the surface then detonated (rather like the HESH shell used by tanks). The results are admittedly impressive, as it can blow its way through 200mm (7.8in) of reinforced concrete. The capabilities of the RAW are somewhat lessened by the awkwardness of its shape and attachment. It is undoubtedly useful for short-range urban combat roles and its explosive power is greater than the M203, but whether it will become a regular attachment on the end of rifles remains to be seen.

| | |
|---|---|
| Country of origin: | USA |
| Calibre: | 140mm |
| Length: | 305mm (12in) |
| Weight: | 3.8kg (8.36lb) |
| Barrel: | Not applicable |
| Feed/magazine capacity: | Single round |
| Operation: | Rifle fired |
| Muzzle velocity: | 180mps (590fps) |
| Effective range: | 200m (656ft) plus |

# Glossary

**Battery** – Descriptive term for when a cartridge is in place and the gun is ready for firing.

**Bolt** – The part of a firearm which usually contains the firing pin or striker and which closes the breech ready for firing.

**Blowback** – Operating system in which the bolt is not locked to the breech, thus it is consequently pushed back by breech pressure on firing and cycles the gun.

**Breech** – The rear of the gun barrel.

**Breech-block** – Another method of closing the breech which generally involves a substantial rectangular block rather than a cylindrical bolt.

**Bullpup** – Term for when the receiver of a gun is actually set in the butt behind the trigger group, thus allowing for a full length barrel.

**Carbine** – A shortened rifle for specific assault roles.

**Chamber** – The section at the end of the barrel which receives and seats the cartridge ready for firing.

**Closed Bolt** – A mechanical system in which the bolt is closed up to the cartridge before the trigger is pulled. This allows greater stability through reducing the forward motion of parts on firing.

**Compensator** – A muzzle attachment which controls the direction of gas expanding from the weapon and thus resists muzzle climb or swing during automatic fire.

**Delayed Blowback** – A delay mechanically imposed on a blowback system to allow pressures in the breech to drop to safe levels before breech opening.

**Double action** – Relates to pistols which can be fired both by cocking the hammer and then pulling the trigger, and by a single long pull on the trigger which performs both cocking and firing actions.

**Flechette** – An bolt-like projectile which is smaller than the gun's calibre and requires a sabot to fit it to the barrel. Achieves very high velocities.

**Gas Operation** – Operating system in which a gun is cycled by gas being bled off from the barrel and used against a piston or the bolt to drive the bolt backwards and cycle the gun for the next round.

**GPMG** – An abbreviation for General Purpose Machine Gun.

**LMG** – An abbreviation for Light Machine Gun.

**Locking** – Describes the various methods by which the bolt or breech block is locked behind the chamber ready for firing.

**Long Recoil** – A method of recoil operation in which the barrel and bolt recoil for a length greater than that of the entire cartridge, during which extraction and loading are performed.

**Muzzle Brake** – A muzzle attachment which diverts muzzle blast sideways and thus reduces overall recoil.

**Open Bolt** – A mechanical system in which the bolt is kept at a distance from the cartridge before the trigger is pulled. This allows for better cooling of the weapon between shots.

**Receiver** – The body of the weapon which contains the gun's main operating parts.

**Recoil** – The rearward force generated by the explosive power of a projectile being fired.

**Recoil Operated** – Operating system in which the gun is cycled by the recoil-propelled force of both barrel and bolt when the weapon is fired. Both components recoil together for a certain distance before the barrel stops and the bolt continues backwards to perform reloading and rechambering.

**SAW** – Abbreviation for Squad Automatic Weapon.

**Self-loading** – Operating system in which one pull of the trigger allows the gun to fires and reload in a single action.

**Short Recoil** – A compressed version of recoil operation in which the barrel and bolt move back less than the length of the cartridge before the bolt detaches.

# Index

**Note:** Page numbers in **bold** refer to main entries.